T5-AFU-553

Production and Logistics

Titles in this Series

Erwin Pesch
Learning in Automated Manufacturing
1994. ISBN 3-7908-0792-3
(Out of print)

Rainer Kolisch
Project Scheduling under Resource Constraints
1995. ISBN 3-7908-0829-6

Armin Scholl
Balancing and Sequencing of Assembly Lines
1995. ISBN 3-7908-0881-4

Dirk C. Mattfeld
Evolutionary Search and the Job Shop
1996. ISBN 3-7908-0917-9

Alf Kimms

Multi-Level Lot Sizing and Scheduling

Methods for Capacitated, Dynamic, and Deterministic Models

With 22 Figures
and 155 Tables

Physica-Verlag
A Springer-Verlag Company

Dr. Alf Kimms
Lehrstuhl für Produktion und Logistik
Institut für Betriebswirtschaftslehre
Christian-Albrechts-Universität zu Kiel
Olshausenstraße 40
D-24118 Kiel, Germany

ISBN 3-7908-0967-5 Physica-Verlag Heidelberg

Die Deutsche Bibliothek - CIP-Einheitsaufnahme

Kimms, Alf:
Multi-level lot sizing and scheduling: methods for capacitated, dynamic, and deterministic models; with 155 tables / Alf Kimms. - Heidelberg: Physica-Verl., 1997
(Production and logistics)
Zugl.: Kiel, Univ., Diss.
ISBN 3-7908-0967-5

SPIN 10548555 88/2202-543210 - Printed on acid-free paper

To my parents
Rita and Werner Kimms

Foreword

This book is the outcome of my research in the field of multi-level lot sizing and scheduling which started in May 1993 at the Christian–Albrechts–University of Kiel (Germany). During this time I discovered more and more interesting aspects about this subject and I had to learn that not every promising idea can be thoroughly evaluated by one person alone. Nevertheless, I am now in the position to present some results which are supposed to be useful for future endeavors.

Since April 1995 the work was done with partial support from the research project no. Dr 170/4–1 from the "Deutsche Forschungsgemeinschaft" (DFG).

The remaining space in this preface shall be dedicated to those who gave me valuable support:

First, let me express my deep gratitude towards my thesis advisor *Prof. Dr. Andreas Drexl*. He certainly is a very outstanding advisor. Without his steady suggestions, this work would not have come that far. Despite his scarce time capacities, he never rejected proof–reading draft versions of working papers, and he was always willing to discuss new ideas — the good as well as the bad ones.

He and *Prof. Dr. Gerd Hansen* refereed this thesis. I am indebted to both for their assessment.

I am also owing something to *Dr. Knut Haase*. Since we almost never had the same opinion when discussing certain lot sizing aspects, his comments and criticism gave stimulating input.

Dr. Rainer Kolisch pointed to some project scheduling literature. The girls in our department's office, *Ethel Fritz*, *Ina Kantowski*, and *Sabine Otte*, provided helpful support. *Uwe Penke* and *Stefan Wende* maintained the computing machinery. *Steffen Wernert* spent hours in front of the copier. He also made the specification of an instance generator become a well–functioning C–program.

Last but not least, I wish to thank *Ulrike.* While this work went on she made sure that production planning was not the only thing in my life, and she forced me to take a breather from time to time.

Kiel, in the spring of 1996 Alf Kimms

Contents

List of Symbols

a_{ji}	"Gozinto"–factor which is defined as the quantity of item j that is directly needed to produce one item i.
$\hat{a}_{ji}$	Entry in an incidence matrix.
α	Encoding function of a path in a gozinto–tree.
ARC_{max}	Parameter to generate gozinto–structures.
ARC_{min}	Parameter to generate gozinto–structures.
$ARCNUM$	Parameter to control the redirection of disjunctive arcs.
$ARCSELECT$	Parameter to control the redirection of disjunctive arcs.
$\mathcal{C}$	Complexity of a gozinto–structure.
C_{mt}	Available capacity of machine m in period t.
$\tilde{C}_{mt}$	Available capacity of the partially renewable resource m in interval t.
$CAPNEED_{max}$	Parameter to generate capacity demand.
$CAPNEED_{min}$	Parameter to generate capacity demand.
CD_{jt}	Cumulative demand for item j in period t.
$CHILD$	Number of child individuals.
$COMPLEXITY$	Parameter to generate gozinto–structures.
$construct$	Program to construct a production plan.
$\Delta costs_{(h,k)}$	Estimated changes of the objective function value when redirecting arc (h,k).
$COSTRATIO$	Parameter to generate setup costs.

$CRITICAL$	Parameter to control the tuning of method parameters.
χ_1	Auxiliary function.
χ_2	Auxiliary function.
d_{jt}	External demand for item j in period t.
$\tilde{d}_{jt}$	Demand matrix entry for item j in period t.
d_{jt}^{L4L}	Production quantities following a lot–for–lot policy.
$\mathcal{DA}$	Set of disjunctive arcs.
$deadline$	Retrieval function.
$deadline^{L4L}$	Retrieval function.
DEM_{max}	Parameter to generate external demand.
DEM_{min}	Parameter to generate external demand.
$demand$	Retrieval function.
dep_j	Depth of item j.
$DEPTH$	Parameter to generate gozinto–structures.
$\mathcal{DS}_{mt}$	Set of node labels which represent item machine m may be set up for at the end of period t.
$fitness_k$	Fitness of individual k.
φ_{mt}	Probability function for choosing a certain setup state.
Γ_j	Gozinto–tree with item j as its root node.
Γ^{DS}	Data structure that is used for demand shuffle.
$\tilde{\Gamma}_j^{DS}$	Representation of an item j in the data structure that is used for demand shuffle.
Γ^{TS}	Data structure that is used for tabu search.
$\tilde{\Gamma}_j^{TS}$	Representation of an item j in the data structure that is used for tabu search.
h_j	Non–negative holding cost for having one unit of item j one period in inventory.
$HCOST_{max}$	Parameter to generate holding costs.
$HCOST_{min}$	Parameter to generate holding costs.
I_{j0}	Initial inventory for item j.

I_{jt}	Inventory for item j at the end of period t.
$\mathcal{I}_{mt}$	Set of items to choose the setup state of machine m at the end of period t among.
id_{ji}	Internal demand for item i caused by producing one item j.
$INITINV_{max}$	Parameter to generate initial inventory.
$INITINV_{min}$	Parameter to generate initial inventory.
$IRAND$	Function to draw an integral random number.
$item$	Retrieval function.
$ITEMS(llc)$	Number of items with low level code llc.
$ITEMSET_{mt}$	Set of items to choose the setup state of machine m at the end of period t among.
j_{mt}	The item machine m is set up for at the end of period t.
J	Number of items.
$\mathcal{J}_m$	Set of all items that share the machine m.
$\tilde{\mathcal{J}}_m$	Set of all items that share the partially renewable resource m.
$\tilde{L}$	Length of an interval in number of periods.
$\mathcal{L}$	Set of all node labels.
$LISTLENGTH$	Length of the tabu list.
LB	Lower bound.
$LEADTIME_{max}$	Parameter to generate lead times.
$LEADTIME_{min}$	Parameter to generate lead times.
$leftwing$	Function to compute a neighbor node.
llc_j	Low level code of item j.
$\mathcal{LW}$	Set of neighbor nodes.
$\lambda_{mt}^{(k)}$	Lagrangean multiplier.
m_j	Machine on which item j is produced.
M	Number of machines.
$\tilde{M}$	Number of partially renewable resources.
$\mathcal{M}_j$	Set of all machines that are capable to produce item j.
$\tilde{\mathcal{M}}_j$	Set of all partially renewable resources

	that are needed to produce item j.
$MACHPERITEM$	Parameter to generate resource assignments.
$mask_{jt}$	Production mask.
$MAXGAP$	Parameter to terminate a lower bounding procedure.
$MAXITER$	Parameter to terminate a lower bounding procedure.
$MAXITER_G$	Parameter to generate gozinto–structures.
$MINLAMBDA$	Parameter to terminate a lower bounding procedure.
$MUTATION$	Mutation probability.
N_C	Parameter to generate costs.
N_D	Parameter to generate external demand.
N_G	Parameter to generate gozinto–structures.
$\mathcal{NL}$	Set of node labels.
$nodes_j$	Number of nodes in the gozinto–tree with item j as its root node.
$nodes_{ji}$	Number of nodes for item i in the gozinto–tree with item j as its root node.
$NODESELECT$	Parameter to control the choice of setup states.
$NOINTENSIFY$	Parameter to control the tuning of method parameters.
nr_j	Net requirement of item j.
OPT	Optimum objective function value.
ω_1	Function to compute a node label.
ω_2	Function to compute a node label.
p_j	Capacity needs for producing one unit of item j.
p_{jm}	Capacity needs for producing one unit of item j on machine m.
$\tilde{p}_{jm}$	Capacity needs for the partially renewable resource m for producing one unit of item j.
$\mathcal{P}_j$	Set of immediate predecessors of item j.
$\bar{\mathcal{P}}_j$	Set of all predecessors of item j.

$PARENT$	Number of parent individuals.
$path$	Retrieval function.
$preds$	Retrieval function.
$PRIVAL$	Parameter to generate gozinto–structures.
π_{jt}	Priority value for setting a machine up for item j at the end of period t.
q_{jt}	Production quantity for item j in period t.
q_{jmt}	Production quantity for item j on machine m in period t.
q_{jt}^B	Production quantity for item j in period t where the required resources are properly set up at the beginning of period t.
q_{jt}^E	Production quantity for item j in period t where the required resources are properly set up at the end of period t.
Q_{jt}	Upper bound for the production quantity of item j in period t.
$RATIODEV$	Parameter to generate setup costs.
RC_{mt}	Remaining capacity of machine m in period t.
$rightwing$	Function to compute a neighbor node.
$RRAND$	Function to draw a real–valued random number.
$\mathcal{RW}$	Set of neighbor nodes.
ϱ_{jt}	Modified priority value for setting a machine up for item j at the end of period t.
s_j	Non–negative setup cost for item j.
s_{jm}	Non–negative setup cost for item j on machine m.
$\mathcal{S}_j$	Set of immediate successors of item j.
$\bar{\mathcal{S}}_j$	Set of all successors of item j.
$SHIFTOPS$	Number of shift operations to be performed per iteration.
$succ$	Retrieval function.
σ	Function to determine the reference counter of a node.

Δt_{left}	Distance of a full–size left shift.
Δt_{right}	Distance of a full–size right shift.
T	Number of periods.
ΔT	Rescheduling interval.
T_{idle}	Number of idle periods.
T_{macro}	Number of macro periods.
T_{micro}	Number of micro periods.
$\tilde{T}$	Number of intervals.
$TIMELIMIT$	Parameter to terminate a lower bounding procedure.
$TYPE_D$	Parameter to generate external demand.
$TYPE_G$	Parameter to generate gozinto–structures.
Θ	Set of setup state selection rules.
U	Capacity utilization.
$\tilde{U}$	Capacity utilization of the partially renewable resources.
UB	Upper bound.
$uplsp$	Program to solve the U–PLSP optimally.
v_j	Positive and integral lead time of item j.
$\mathcal{VC}$	Set of disjunctive arcs which may be redirected.
$\mathcal{VL}$	Set of nodes which may be shifted.
$weight_{(h,k)}$	Weight of the arc (h,k).
x_{jt}	Binary variable which indicates whether a setup for item j occurs in period t ($x_{jt}=1$) or not ($x_{jt}=0$).
x_{jmt}	Binary variable which indicates whether a setup for item j occurs on machine m in period t ($x_{jmt}=1$) or not ($x_{jmt}=0$).
Ξ	Postprocessing function.
y_{j0}	Unique initial setup state.
y_{jt}	Binary variable which indicates whether machine m_j is set up for item j at the end of period t ($y_{jt}=1$) or not ($y_{jt}=0$).
y_{jm0}	Initial setup state.
y_{jmt}	Binary variable which indicates whether

	machine m is set up for item j at the end of period t ($y_{jmt} = 1$) or not ($y_{jmt} = 0$).
$z_{jt\tau}$	Fraction of the gross demand $d_{j\tau}^{L4L}$ for item j produced in period t.

List of Abbreviations

APCIG	a parameter controlled instance generator
B&B	branch–and–bound
BOM	bill of material
CLSP	capacitated lot sizing problem
CPU	central processing unit
CSLP	continuous setup lot sizing problem
DLSP	discrete lot sizing and scheduling problem
ELSP	economic lot sizing problem
EOQ	economic order quantity
GRASP	greedy randomized adaptive search procedure
LP	linear program
MIP	mixed–integer program
MPS	master production schedule
MRP	material requirements planning
MRP II	manufacturing resource planning
PLSP	proportional lot sizing and scheduling
PLSP–MM	proportional lot sizing and scheduling with multiple machines
PLSP–MR	proportional lot sizing and scheudling with multiple resources
PLSP–PM	proportional lot sizing and scheduling with parallel machines
PLSP–PRR	proportional lot sizing and scheduling with partially renewable resources
RCPSP	resource constrained project scheduling problem
U–PLSP	PLSP–MM without capacity constraints
WSP	warehouse scheduling problem

Chapter 1

Introduction

Lot sizing[1] certainly belongs to the most established production planning problems.[2] As first scientific reports of this subject date from the beginning of the 20th century [And29, Har13] and at least one chapter about lot sizing can be found in almost every good textbook about production research issues, it seems to be hard to justify yet another publication focusing on lot sizing. Especially a monograph appears to be of no further relevance at first glance.

But, as we will show, the state–of–the–art does not provide any satisfying solutions to some questions of practical importance such as multi–level lot sizing, especially not, when capacity is scarce and demand is time variant. Furthermore, lot size and sequence decisions are usually not integrated as it ought to be for the short–term planning as argued in [PoWa92].

This book fills in the gap and presents models and methods for multi–level lot sizing and scheduling. The first chapter introduces the reader into the planning problem. It illustrates current practice

---

[1] A *lot size* is defined as "The amount of a particular item that is ordered from the plant or a supplier or issued as a standard quantity to the production process" (see [APICS95a]).

[2] It is worth to be mentioned that lot sizing is not a production planning problem only. In cash management, for instance, the transactions demand for cash — as motivated by Keynes — implies a cash balancing problem. On the one hand there is the risk of cash insolvency, while on the other hand there is the value of forgone interest [BrMy88]. This trade–off can be modelled as a lot sizing problem [Bau52], too.

and motivates what is to come subsequently. Eventually, a chapter overview guides the reader.

1.1 Problem Context

Consider the organization of an in–house production system. Typically, the architecture of such a system is build up from several production cells, so–called segments, which may be implemented in different fashions (flow lines or work centers for instance). This macro–structure further refines into a micro–structure as each segment provides the capability to perform a bunch of operations.

Raw materials and component parts are floating concurrently through this complex system in order to be processed and assembled until a final product comes out being ready for deliverance.

Production planning is one of the most challenging subjects for the management there. It appears to be a hierarchical process ranging from long– to medium– to short–term decisions (see for instance [DrFlGüStTe94, Fle88, Gün86, SöSc95, Sta88a, Swi89, Vil89]). Our focus will be the short–term scope which links to medium–term decisions via the master production schedule (MPS). The MPS defines the external (or independent) demand, i.e. due dates and order sizes for final products. The goal now is to find a feasible production plan which meets the requests and provides release dates and amounts for all products including component parts. For economical reasons, finding a feasible plan is not sufficient. In the usual case, production plans can be evaluated by means of an objective function (e.g. a function which measures the setup and the holding costs). Then, the aim is to find a feasible production plan with optimum (or close to optimum) objective function value.

1.2 Problem Outline and General Issues

Let the manufacturing process be triggered by orders which originate from customers or from other facilities. Suppose now, that the output of the *make–to–order* system under concern is or at least includes a set of *non–customized products*. Certainly, this is a valid assumption for many firms no matter what industry they belong to and no matter of what size they are.

To motivate a planning activity, we first need to identify a subject of concern that is worth (in terms of economical rationale) to be considered. A first clue are large inventories. Due to the opportunity costs of capital and the direct costs of storing goods, holding items in inventory and thus causing holding costs should be avoided. On the other hand, if different parts are making use of common resources, say machines, and a setup action must take place to prepare proper operation, then opportunity costs (i.e. setup costs) are incurred since production is delayed. Another aspect of sharing resources is that the production of such parts cannot coincide if different setup states are required. Hence, orders must be sequenced. If production planning is about the timing of production and not about what to produce (e.g. make–or–buy decisions), then production costs need not to be considered as long as they are time invariant. In summary, we have a trade–off between low setup costs (favoring large production lots) and low holding costs (favoring a lot–for–lot–like production where sequence decisions have to be made due to sharing common resources). Essentially, the problem of short–term production planning turns out to be a *lot sizing and scheduling problem* then.

If we ask about how to solve this production planning problem, we first need a deeper understanding of its basic attributes. The first key element we have to remember is the stream of component parts floating through a complex production system. Operations may be executed only if parts being subject of these particular operations are indeed available. In other words, a production plan must respect the precedence relations of operations. Hence, *multi–level* structures must be taken into account. For the sake of convenience, we do not further distinguish between operations and items (also called products or parts). Each operation produces an item, and each item is the output of an operation. Both terms are used as synonyms (see also [Tem95]). Apparently, we face a *multi–item* problem here.

The second key element of our problem is the presence of scarce capacity. As usual in in–house production systems, producing an item requires a certain amount of one or more resources (e.g. manpower, machine time, energy, ...) with limited capacity per time unit. Thus, production planning must take *scarce capacity* into account. Following the terminology in the literature, we also say machine instead of resource.

Furthermore, we have the following situation: The time interval which is the focus of the planning process is *finite* and subdivided into several *discrete time periods.* To refer to these periods we number them consecutively beginning with period 1. The length of the overall time interval is called the planning horizon, or horizon for short, and is counted in number of time periods. This properly reflects the real–world situation where we face a planning horizon of say four weeks (or 20 days or 40 shifts) and discrete time periods are naturally given. The (known or estimated) external demand (given by the MPS) is given in units per item per period. It is to be met promptly at the end of each period. Backlogging and shortages are not allowed here which enforces a high service level. The demand may vary from period to period. This is called *dynamic* demand. All relevant data for the planning process is assumed to be *deterministic* which is justified by having a short–term planning problem on hand.

1.3 Case Descriptions

To underscore the practical importance of multi–level lot sizing and scheduling, let us have a look at real–world situations demanding for methods to be applied. Two cases will be briefly described: Eastman Kodak Company (an elaborate analysis and results of a simulation of this case can be found in [KaKeKeFr85]) and Owens–Corning Fiberglas Corporation (this case is described in [OlBu85]). Other real–world examples do, of course, exist. E.g. not reported about here is production planning at Lever Sunlicht GmbH, a washing powder producer, which is informally analyzed in [Eif75]. Mathematical models of cases can be found in [Gor70] (tire production) and [ViMa86] (pharmaceutical industry).

1.3.1 Eastman Kodak Company

Eastman Kodak's Apparatus Division, Rochester, New York, USA, has implemented a manufacturing cell where 13 similar parts are produced. These parts were grouped because of long production lead times, high in–process inventories, and difficulties in coordinating assemblies. The cell contains 10 major work centers plus three minor facilities for preparation and finishing operations. Seven out of the 10 major work centers have one machine only, while the remaining three work centers have more than one machine each. The machines in the cell are manual and numeric controlled lathes, drills, punches and other metal forming processing units. Some of them are identified as bottlenecks. The parts do not flow uniformly through the cell. Subject of concern are lot sizing policies for the cell.

1.3.2 Owens–Corning Fiberglas Corporation

Owens–Corning Fiberglas, Anderson, South Carolina, USA, produces a multitude of fiber glass products, among them over 200 mat items as they are needed for boat hulls, pipeline construction or bathroom fixtures. 28 of the mat items are non–customized products, and 80 percent of the external demand is for these standard mats. Mats are sold in rolls in various widths and weights, treated with one out of three binders, and trimmed on one or both edges or not at all. Two parallel production lines with different capacities and speed are available. On one line, mats with up to 76 inches width can be produced, while on the other, up to 60–inch material can be manufactured. The former one works about three times faster than the latter. Both lines perform the same set of operations. Raw fiber glass is pulled through high–speed choppers first. Then, these bits and pieces are spread over a mat chain via a forming hood to be passed on to binder applicators. In the next step, the mat is put into a drying oven. Afterwards, it is carried to a compaction roller to add strength. The following production stage is trimming. Finally, the mats are rolled into 175– to 230–pound cylinders. A careful analysis of the production process revealed that at each level maintenance costs are closely related to the frequency of setup operations. Also, significant costs for downtime and expensive mat waste result from each setup. The objective is to find line assignments, lot sizes, and sequences.

1.4 Current Practice and Motivation

The basic working principle of today's decision support systems for Manufacturing Resource Planning (MRP II) is more or less the same in all current implementations. It should be reviewed here by means of a small example (see also [DrHaKi95]). For an overview of more than a hundred modern MRP II software packages we refer to [APICS95b, APICS95c] where a detailed list of features is enclosed.

Assume the following data: Three items $j = 1, \ldots, 3$ are to be produced sharing a single machine. The gozinto–structure[3] of these items is given in Figure 1.1. The planning horizon is 10 periods ($t = 1, \ldots, 10$) long and inventory is empty. Table 1.1[4] provides the MPS (with external demands d_{jt}), the capacity limit per period C_t, the need of capacity per item p_j, and the item–specific setup costs s_j and holding costs per period h_j. We assume that the minimum lead time is zero.

Starting with the MPS, lot sizes for all items are determined by a level–by–level approach disregarding capacity constraints. In our example, we start to compute lot sizes for item 1 using some lot sizing rule. Let us say, we decide to use a lot–for–lot policy. Then, we derive internal demands and due dates for the next level which is item 2. As a result we have an internal demand for 40 units of item 2 in periods 6, 8 and 10. Again, we employ some lot sizing rule, but, this time considering item 2. Suppose, we decide

[3]The term gozinto–structure was coined by Vazsonyi [Vaz58] who gave a reference to an italian mathematician named Zepartzat Gozinto. This person is pure fiction. But, if you read his name in English pronunciation, it turns out what is meant: The product–structure. Nodes represent items, and arcs depict precedence relations between the items. Arc weights are production coefficients.

[4]Missing entries in tables are assumed to be zero throughout the text.

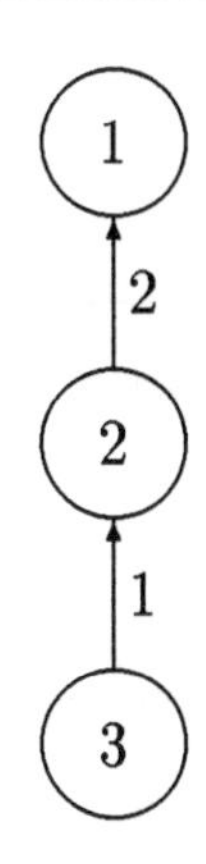

Figure 1.1: A Gozinto–Structure with Three Items

d_{jt}	$t = 1$	...	6	7	8	9	10	h_j	s_j	p_j
$j = 1$			20		20		20	25	900	1
$j = 2$								10	850	1
$j = 3$								10	800	1
C_t	100	...	100	100	100	100	100			

Table 1.1: Data of the Example

to produce the demands of periods 8 and 10 in one lot in period 8. Following the same lines, we then compute internal demands for item 3 and again employ some lot sizing rule. As a result we might get what is shown in Table 1.2 where q_{jt} denotes the production quantity of item j in period t. Note, this is not a valid production plan since capacity restrictions are violated in period 8.

In a next step, the intermediate result is modified to find a plan without requiring an excess of capacity. This is done by shifting lots to the left or to the right until the capacity profile is met. In our example for instance, this can be achieved by shifting the lot for item 2 in period 8 to the left (see Table 1.3).

q_{jt}	$t=1$	2	3	4	5	6	7	8	9	10
$j=1$						20		20		20
$j=2$						40		80		
$j=3$						40		80		

Table 1.2: A Production Plan with Capacity Constraint Violations

q_{jt}	$t=1$	2	3	4	5	6	7	8	9	10
$j=1$						20		20		20
$j=2$						40	80			
$j=3$						40		80		

Table 1.3: A Production Plan with Precedence Relation Violations

But, because precedence constraints are not taken into account when lots are shifted in order to find a plan that does not violate the capacity restrictions, the conventional approach obviously fails. Looking at the example reveals, it would never be possible to produce item 2 in period 7 since we lack a sufficient amount of item 3 there.

The traditional way to overcome this is to introduce lead times. To make this idea clear, let us start with the MPS again (see Table 1.1). Going through the level–by–level approach as described above, we now use an offset of, say, two periods — the so–called lead time — when we compute the due dates of the internal demand. The outcome of the procedure is given in Table 1.4.

q_{jt}	$t=1$	2	3	4	5	6	7	8	9	10
$j=1$						20		20		20
$j=2$				40		80				
$j=3$		40		80						

Table 1.4: A Production Plan with Positive Lead Times

Note, this plan is infeasible again. Once more, we are off capacity limits. Shifting the lot of item 2 in period 4 to the left, leads us to the solution in Table 1.5.

q_{jt}	$t=1$	2	3	4	5	6	7	8	9	10
$j=1$						20		20		20
$j=2$			40			80				
$j=3$		40		80						

Table 1.5: A Feasible Production Plan

Now we are done. The result represents a feasible production plan. The sum of setup and holding costs is 9,800. Due to the introduction of positive lead times we had more flexibility when shifting lots which helped. And, indeed this is what can be observed in practice. Known as the lead–time–syndrome [ZäMi93], planners tend to increase lead times arbitrarily whenever they detect backorders or high work–in–process inventories. But, as a result most firms suffer from long makespans and large total holding costs. To be convinced, compare the feasible plan in Table 1.5 with the optimum plan in Table 1.6. Figure 1.2 shows the schedule of the optimum solution by means of a Gantt–chart. The sum of setup and holding costs is 6,700.

q_{jt}	$t=1$	2	3	4	5	6	7	8	9	10
$j=1$						20		20		20
$j=2$						40		80		
$j=3$						40	80			

Table 1.6: An Optimum Production Plan

The reason for this dilemma in production planning apparently is the traditional level–by–level approach. This widely used method does not take capacity constraints and precedence relations

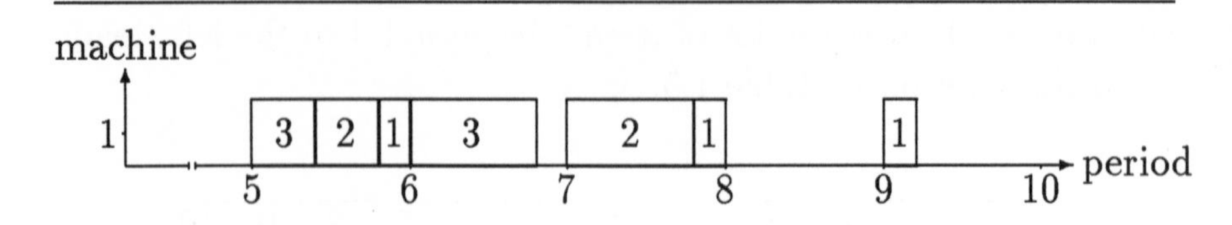

Figure 1.2: An Optimum Schedule

simultaneously into account. Unfortunately, good alternatives are not available yet. This gives the motivation to develop such.

1.5 Chapter Synopsis

In this book we find 13 chapters of presentation of research on multi–level lot sizing and scheduling. An introduction plus a motivation for spending effort to this subject is given in Chapter 1. Chapter 2 gives a review of the relevant literature to prove that this contribution is indeed new and challenging. Formal specifications of the problems under concern are given in Chapter 3. Chapter 4 introduces what is needed for the computational studies. Lower bounds are given in Chapter 5. Then, in Chapter 6, several methods for solving multi–level lot sizing and scheduling problems with multiple machines are presented. Chapter 7 investigates the extension with parallel machines. So does Chapter 8 with multiple resources and Chapter 9 with partially renewable resources. Chapter 10 is about rolling horizon implementations. The question of what method to use in what situation is answered in Chapter 11. Chapter 12 gives an overview of research opportunities. Finally, in Chapter 13 concluding remarks finish this contribution. The rest of this section is good for giving a little more detail of the contents of the subsequent chapters.

Chapter 2 contains the most comprehensive literature review of multi–level lot sizing available today. Though contributions assuming continuous demand are of no help for the dynamic case they are enumerated for the sake of completeness. Details are given for those references only which consider dynamic demand. As a result we will find out that this book could indeed make a significant contribution because we discover that multi–level lot sizing and scheduling is an untreated topic.

In **Chapter 3** we give formal specifications of the problems under concern by means of mixed–integer programs (MIPs). As we will show, the presented models are new in the sense that they make much more general and therefore more realistic assumptions than other well–known models. Hence, many widespread models are special (and thus easier) cases. The most basic model assumes multiple machines. The parallel machine case, which is almost untreated even under the single–level assumption, is specified as well. Besides this, we extend the multiple machine model to a model for so–called multiple resources. Furthermore, we introduce a new notion into the lot sizing world, namely partially renewable resources. The chapter finishes with some first problem insights which give a feeling for its complexity and hardness.

There is no standard test–bed available which could be used to evaluate new methods. For this reason we define an instance generator in **Chapter 4**. It can then be used to construct a factorial, experimental design. Some remarks about the performance of computers will later help to assess the run–time performance of methods.

Standard solvers are not capable to solve medium– to large–size problem instances. To evaluate the performance of methods for such sizes, we thus need some lower bounds. These are derived in **Chapter 5**. Beside LP–relaxations of the original model and a MIP–reformulation, we also relax capacity constraints and solve the remaining problem optimally with a branch–and–bound (B&B) procedure. Additionally, a Lagrangean relaxation of the capacity restrictions solved on the basis of the B&B–method is evaluated.

In **Chapter 6**, which is the main part of this book, several methods for multi–level lot sizing and scheduling with multiple machines are developed and tested. Among them are randomized regret based sampling, tabu search with disjunctive arc based problem representations, and genetic algorithms. Furthermore, we introduce the idea of cellular automata which has never before been applied to optimization problems. In addition to that, a novel so–called demand shuffle heuristic which is based on random sampling combined with problem specific data structure manipulations is invented. Unsuccessful attempts are reported as well.

Chapter 7 is about the parallel machine case. A modification of the demand shuffle procedure is discussed and preliminary computational tests are performed. Since lot sizing with parallel machines is quite a new challenge, we review the job scheduling literature for related work.

In **Chapter 8** we apply a variant of the demand shuffle approach to the multiple resource extension, and again do some preliminary tests. For related topics, we review the project scheduling literature.

We examine yet another modification of the demand shuffle method for partially renewable resources in **Chapter 9**. The notion of partially renewable resources is new, especially in the area of lot sizing.

Since production planning is usually done on a rolling horizon basis, problems related to it are discussed in **Chapter 10**. With a review of the literature we also get acquainted with nervousness, stability measures, and rescheduling.

Chapter 11 discusses the question of how to choose the right method given a particular instance.

In **Chapter 12** we give some hints for future work. Beside some straightforward extensions, an emphasis is on the interdependencies with other planning tasks in a firm.

Finally, in **Chapter 13** we conclude the work.

A comprehensive list of references completes the book.

Chapter 2

Literature Review

This chapter contains the most comprehensive review of the multi-level lot sizing literature available today. First, in Section 2.1 the scope of the review is defined. In Section 2.2 we then list some contributions of general interest. Models and methods for dynamic demand and no capacity constraints are discussed in Section 2.3. In Section 2.4 our concern is about dynamic demand and capacity restrictions. In both sections we refer to methods, e.g. dynamic programming or simulated annealing without further comment, because their basic working principles are more or less standard. Readers who are not familiar with these may start with [ZaEvVa89] for an overview of heuristics and with any good operations research textbook (e.g. [DoDr95, HiLi88, Win93]) for exact methods. Finally, in Section 2.5 intermediate conclusions from the review are given.

2.1 Scope of the Review

The huge amount of literature can be classified into several categories. To separate those contributions which are of relevance for our research from those which are not, we discriminate three important criteria.

First, publications may assume stochastic or deterministic data. The focus here is on deterministic cases. Since models and methods under stochastic assumptions differ very much from those which assume deterministic data, the former ones are of no interest here. We refer to [AxRo94, ChZh94, GoLeGaXi94, GrMeDaQu86, Jah95, YaLe95] for an introduction.

Second, authors assume a single– or a multi–level production. Approaches of the former type do not fit for multi–level structures by definition. Hence, we do not give a comprehensive overview of single–level lot sizing here. [AkEr88, BaRiGu87, Bak90, DoScVo93, Elm78, Fle88, KuSaWa94, Tem95] contain reviews. However, modifications of single–level heuristics may be used to attack multi–level problems, e.g. on a level–by–level basis. Such methods are called improved heuristics. Subsequent sections deal with these.

Last, demand may be assumed to be dynamic or stationary. In a stationary situation, the demand occurs continuously and with a constant rate. Usually, an infinite planning horizon comes along with stationary demand. This is appropriate for mass production for instance. In most cases, the common basis is an economic order quantity (EOQ) policy [And29, Har13] with no capacity considerations, or the economic lot sizing problem (ELSP) [Elm78] where capacity limits are taken into account. There is a vast amount of literature about multi–level lot sizing for stationary demand. Most of it is based on EOQ–like assumptions. But, these approaches are of minor help in the dynamic case. Hence, it is sufficient to cite some work without going into the details: [AtQuSu92, AtSu95, BeBrPu85, BiMo79, Cha84, ChSh85, Cla72, CoVa82, CrWaHe72, CrWaWi73, ENa89, ENa92, ENaKl93, ENa94, GaJo94, Goy76, GrSc77, GrSc78, HvL67, HaYa95a, HaYa95b, Hof94, HsENa90, Hsu84, HuSa91, IyAt93, JaMaMu88, JeKa72, MaMu85, MoBi77, MClTr85, Moi82, Moi86, MuSi78, MüMe65, Rou86, Rou89, Sch68, Sch73, ScSc75, ScSc78, Sze75, Sze76, Sze78, SzDr80, Sze81, Sze83, TaSk70, Wil81, Wil82].

In the subsequent sections we should have a closer look at the multi–level lot sizing literature where data are deterministic and demand is dynamic. Though our interest lies in capacitated situations primarily, uncapacitated cases are closely related and shall therefore be reviewed, too.

2.2 Contributions of General Interest

This is of course not the first review of (multi-level) lot sizing. So, let us start with a look at the history of surveys. A very brief survey was given by Jacobs and Khumawala [JaKh80] with seven references only. Research issues for multi-level lot sizing are discussed by Collier in [Col82b]. DeBoth et al. [DBoGeWa84] review dynamic lot sizing. The review of Bahl et al. [BaRiGu87] probably is the one that is most often referenced. Fleischmann [Fle88] gives insight how operations research methods were applied to different production planning problems. So he does for lot sizing. Gupta and Keung [GuKe90] also provide a review. Goyal and Gunasekaran [GoGu90] provide a collection of literature about multi-level production/inventory systems. Recently, Kuik et al. [KuSaWa94] report on different assumptions for lot sizing, among them multi-level structures. Simpson and Erenguc [SiEr94a] compiled literature related to multi-level lot sizing, too, and conclude: "Clearly, work in this area has only begun." In the textbooks written by Domschke et al. [DoScVo93] and Tempelmeier [Tem95] we also find notable subsections explaining multi-level lot sizing research.

The most basic problem in material requirements planning (MRP) introduced by multi-level structures is the parts requirement (or explosion) problem which is that of computing the total number of parts needed to fulfill demand. In this context, the low level code (or explosion level) of an item is defined as the number of arcs on the longest path from that particular item to an end item. Especially in the presence of positive component inventory levels, the determination of the so-called net requirement has attracted early researchers. Vazsonyi [Vaz58] gives a mathematical statement. Elmaghraby [Elm63] is among the first who present an algorithm. Thompson [Tho65] follows up and coins the notion of a technological order which is a partial order on the basis of the low level codes of the items. Both of them use matrix operations to compute net requirements. Nowadays, this problem does no longer exist since efficient methods using the low level codes of the items are available.

The bill of material (BOM) is a formal statement of the successor–predecessor relationship of items and thus defines the gozinto–structure. Special cases of acyclic structures are those where each item has at most one successor (so–called assembly structures), and those where each item has at most one predecessor (so–called divergent, distribution, or arborescence structures, respectively). Structures which belong to the assembly as well as to the distribution type are called serial (or linear). A gozinto–structure that is neither assembly nor divergent is general. Collier [Col81, Col82a] is the only who tries to define a measure of complexity for gozinto–structures by making use of the information in the BOM. He introduces the degree of commonality index which reflects the average of the number of successors per component part. A formal definition will be given in a later chapter when we compare his index with one of our own.

2.3 Dynamic Demand, Unlimited Capacity

Clark and Scarf [ClSc60] provide an early analysis of optimal lot sizing policies in a distribution network. They also invent the notion echelon stock for what is in inventory plus what is in transit but not sold. Veinott [Vei69] formulates lot sizing as minimizing a concave function over the solution set of a Leontief substitution system. His work is a generalization of Zangwill's paper [Zan66]. For serial gozinto–structures Zangwill [Zan69] presents a dynamic programming approach based on a reinterpretation of the constraints as flow constraints in a single–source network. Love [Lov72] uses these results and considers serial structures under certain assumptions about the costs related to different production levels. He presents a dynamic programming algorithm, too, to find so–called nested extreme optimal schedules where the attribute nested means that production for an item in a certain period implies production for all successor items in the same period. Crowston and Wagner [CrWa73] extend Love's model and develop dynamic programming and B&B–methods for assembly systems. On the basis of Veinott's and Zangwill's results, Kalymon [Kal72] decomposes problems with divergent gozinto–structures into a series of single–item problems. He uses an enumeration procedure to determine the sequence in which these single–item problems are then solved.

Steinberg and Napier [StNa80] determine optimal solutions for general gozinto–structures with standard MIP–solvers on the basis of a network model. The production quantities may be constrained by an item–specific upper bound. An alternative formulation with a less cumbersome notation is given by McClain et al. [MCMMTW82]. In response, Steinberg and Napier [StNa82] make clear that the alternative formulation is more compact, but that there is no evidence which proves greater computational efficiency.

Rao [Rao81] uses Bender's decomposition to compute optimal solutions for problems with general gozinto–structures.

Krajewski et al. [KrRiWo80] perform a computational study and by using its outcome they determine the parameters of three regression models with least squares estimators. For instance, they specify the average size of unreleased orders for each item as a function of the number of product levels above the item, the total number of components for the item, the number of immediate components, and the general lot size for the system (which is a binary variable indicating small or large lots).

Bitran et al. [BiHaHa83] suggest a planning approach where the gozinto–structure is aggregated into just two levels. Then, the two–level problem is solved and the solution is eventually disaggregated. However, they do not consider setups which makes both, the two–level planning problem as well as the disaggregation, become linear with continuous decision variables only.

Bahl and Ritzman [BaRi84] formulate a non–linear mixed–integer program for simultaneous master production scheduling, lot sizing, and capacity requirements planning. They also develop a heuristic. The traditional level–by–level approach is retained. Furthermore, they assume a lot–for–lot policy for all component parts. In fact, they reduce the multi–level problem to a single–level one under these assumptions.

By means of simulation studies Biggs [Big79] examines the

interaction effect of different lot sizing and sequencing rules when used conjunctively. In a first step, he solves a multi–level lot sizing problem using a single–level lot sizing rule on a level–by–level basis. Note that the same rule is applied to all levels. For each period he computes a sequence of different items with a simple priority rule, afterwards. The test–bed consists of six lot sizing rules combined with five sequencing rules. The performance is analyzed using several criteria, among them the total number of setups for instance. As an outcome of this study, it turns out that interaction effects between lot size and sequence decisions do exist, but, "... a total explanation will have to wait for future research...".

Billington et al. [BiMClTh83] discuss how to reduce the size of problem instances by means of gozinto–structure compression. A similar idea is published by Axsäter and Nuttle [AxNu86, AxNu87] where assembly structures are taken into account only. Zangwill [Zan87] analyzes problems with serial gozinto–structures to find out at which levels inventory cannot occur. Though his motivation is to give advice where to invest for setup cost reduction in order to enable just–in–time production and to achieve zero inventories irrespective of demand, the information obtained can also be used to compress the gozinto–structure.

When testing six single–level heuristics on a level–by–level basis using the same lot sizing rule on each level, Benton and Srivastava [BeSr85] find out that neither the depth nor the breadth of a gozinto–structure has a significant effect on the performance of lot sizing. Afentakis [Afe87] conducts a computational study which contradicts this result. The reason might be that Benton and Srivastava use gozinto–structures with up to five levels and six items while Afentakis uses structures with up to 45 levels and 200 items. Sum et al. [SuPnYa93] provide a study which confirms Afentakis. They employ 11 lot sizing rules on a level–by–level basis tested on 1980 instances. By the way, a heuristic that is introduced by Bookbinder and Koch [BoKo90] performs best. An analysis of variance reveals the impact of the number of items, the number of levels, and the average number of immediate successors per item [Col81]. It is interesting to note that the ranking of the rules remains stable.

Simplistic applications of single–level lot sizing rules are also described by Berry [Ber72] and Biggs et al. [Big75, BiGoHa77, BiHaPi80]. Another evaluation of the level–by–level technique is done by Veral and LaForge [VeLFo85]. Choi et al. [ChMaTs88], Collier [Col80a, Col80b], Ho [Ho93], LaForge [LFo82, LFo85], and Lambrecht et al. [LaVEeVa81] provide computational studies, too.

A study in which single–level lot sizing rules are applied to a multi–level structure in a level–by–level manner is done by Yelle [Yel76]. He uses four different simple lot sizing rules and a two–level test–bed. His basic idea is to apply different rules to different levels. This study is extended by Jacobs and Khumawala [JaKh82] who add three more single–level methods plus a multi–level algorithm proposed by McLaren [McLa76] which in turn bases on the single–level Wagner–Whitin procedure [AgPa93, FeTz91, WaHoKo92, WaWh58]. Tests with a three–level test–bed are done by Choi et al. [ChMaCl84].

Blackburn and Millen [BlMi82a] apply single–level heuristics level–by–level to assembly structures using the same rule for all levels. Three heuristics are tested this way. Following the ideas of New [New74] and McLaren [McLa76], they modify the cost parameters of the items in order to take the multi–level structure indirectly into account. Five types of modifications are tested. Closely related to their lot sizing approach is the one of Rehmani and Steinberg [ReSt82] who keep the cost parameters, but, modify the computation of some derived values that guide a single–level heuristic.

Gupta et al. [GuBr92] compare the cost modification approach with others. Studies of Wemmerlöv [Wem82] also examine the impact of cost modifications.

Raturi and Hill [RaHi88] introduce something similar to capacity constraints into the model of Blackburn and Millen. But, capacity usage is only estimated on the basis of average demand. They heuristically compute shadow prices for the capacities and derive modified setup costs from that. Eventually, the lot sizing problem is solved using the Wagner–Whitin procedure level by level.

An extension of Blackburn and Millen's approach for general gozinto–structures is presented by Dagli and Meral [DaMe85]. The combination of lot sizing and capacity planning is treated in [BlMi84] where the subject is confined to assembly structures. The impact of a rolling horizon is discussed in [BlMi82b].

Combining the idea of using different single–level heuristics at different levels with the idea of modifying the cost parameters is evaluated by Blackburn and Millen [BlMi85]. Seven algorithms are used in six combinations. In all tests they choose one algorithm for the end item level and another algorithm for the remaining levels.

While the above level–by–level algorithms construct a solution in one pass, Graves [Gra81] presents a multi–pass procedure. Assuming that external demand occurs for end items only, the basic idea is to perform a single pass, and then to modify the cost parameters (how this is done is of no relevance here). Afterwards, he solves the single–item problems for those items again for which external demand exists. If nothing changes for these items the procedure terminates. If there are any changes in the production plan then the whole procedure repeats using the modified cost parameters this time. A proof of convergence is given.

Another multi–pass heuristic is invented by Peng [Pen85] who uses the Wagner–Whitin procedure to generate an initial solution. This solution is then repeatedly reviewed to combine lots until no further cost savings can be achieved.

Moving away from level–by–level lot sizing, Lambrecht et al. [LaVEeVa83] compare level–by–level approaches with a period–by–period approach. The latter one computes production quantities for the first period and for all items before the second period is concerned. This is going on period by period. All computational tests are restricted to assembly structures. As a result, the period–by–period approach seems to be competitive. A somehow similar method proposes Afentakis [Afe87]. Starting with solving the lot sizing problem consisting of the first period only, he generates a solution which includes the next period by augmenting the intermediate result. This is done until all periods are covered. While Afentakis uses the term sequential approach instead of level–by–level, he now calls his method a parallel one instead of period–by–period. The proposed method decidedly outperforms all (five) single–level methods that are used level–by–level in his study.

Rosling [Ros86] gives an uncapacitated plant location reformulation of the problem with assembly structures. On the basis of this new MIP–model, he develops an optimal procedure.

Heinrich [HeSc86, Hei87] presents a heuristic for general gozinto–structures which operates in two phases. First, the dynamic multi–level lot sizing problem is reduced into a problem with constant demand (using the average demand for each item for instance). Using modified cost parameters, the resulting problem is then solved with a deterministic search in the set of feasible reorder points. In the second phase, the resulting production amounts are modified to meet the demand of the original problem instance.

Coleman and McKnew [CoMKn91] keep the idea of level–by–level methods alive and design a new heuristic based on an earlier work [CoMKn90]. Although their computational study gives promising results, they fail to compete with sophisticated, established methods (e.g. [Gra81]). A study by Simpson and Erenguc [SiEr94b] indicates that, if this is done, the new heuristic appears rather poor.

Gupta and Brennan [GuBr92] evaluate level–by–level approaches if backorders are allowed. Other improved heuristics for models with backlogging are discussed by Vörös and Chand [VöCh92].

Afentakis et al. [AfGaKa84] efficiently compute optimal solutions for assembly structures. The proposed method is of the B&B–type. Based on a MIP–reformulation, lower bounds are determined with a Lagrangean relaxation of those constraints which represent the multi–level structure. As a result, a set of single–item, uncapacitated lot sizing problems is to be solved. This is done with a shortest path algorithm. Optimal solutions for general gozinto–structures are determined by Afentakis and Gavish in [AfGa86] by transforming complex structures into assembly structures where some items occur more than once. Chiu and Lin [ChLi89] also compute optimal solutions for assembly structures with dynamic programming. Their idea is based on a graphical interpretation of Afentakis' reformulation. A level–by–level heuristic with postprocessing operations is described as well.

Kuik and Salomon [KuSa90] apply a stochastic local search method, namely simulated annealing, to multi–level lot sizing. This approach can handle general gozinto–structures. A comparison with other approaches is not done.

Salomon [Sal91] suggests a decomposition of the multi–level problem into several single–item problems. Roughly speaking, this is achieved by eliminating inventory variables by substitution, doing a Lagrangean–like relaxation of inventory balances, and adding some new constraints. He then solves the single–item problems with some lot sizing algorithm sequentially, and upon termination updates the Lagrangean multipliers to repeat the process.

Joneja [Jon91] considers assembly gozinto–structures. By adapting results from lot sizing with stationary demand [Rou86], he develops a so–called cluster algorithm and proves worst case performance bounds. The projection of this work to serial structures closely relates to Zangwill's paper [Zan87]. Roundy [Rou93] presents two cluster algorithms for general structures and worst case bounds for these. In a computational study, one of them turns out to be slightly better than Afentakis' method [Afe87].

McKnew et al. [MKnSaCo91] present a linear zero–one–model for multi–level lot sizing. Assembly gozinto–structures are assumed. They claim that the LP–relaxation of the model always yields integer solutions. Rajagopalan [Raj92] arguments that they are wrong. But, his counter–example is false.[1]

Pochet and Wolsey [PoWo91] consider general gozinto–structures and derive cutting planes to ease the computational effort when standard solvers are running.

Atkins [Atk94] suggests a way to compute lower bounds for problems with assembly structures. His basic idea is to convert an assembly structure into a set of serial structures by finding all paths from the lowest level to end items. Some items may now appear in more than one serial structure having their own identity. In such a case the original setup costs are split up and assigned to

[1] See [Raj92], page 1025: The determinant of the submatrix in the example evaluates to 1 and not to 2 as stated in the paper. Hence, there is no contradiction to total unimodularity.

the (new) items. Eventually, the approach in [Lov72] can be used to solve the resulting instances which, in summary, gives a lower bound.

Simpson and Erenguc [SiEr94b] invent a neighborhood search procedure and slightly outperform [Gra81], [CoMKn91], and three other improved heuristics. Lower bounds are obtained with a Lagrangean relaxation.

Richter and Vörös [RiVö89] are trying to find a so–called setup cost stability region which is defined as the set of all setup cost values for which a given solution remains optimal. Recently, Vörös [Vör95] analyzes the sensitivity of the setup cost parameters for facilities in series and computes stability regions with a dynamic programming procedure.

Arkin et al. [ArJoRo89] give a classification of the complexity of uncapacitated, multi–level lot sizing problems. Optimization problems with general gozinto–structures and no setup times are proven to be NP–hard.

2.4 Dynamic Demand, Scarce Capacity

Lambrecht and Vander Eecken [LaVEe78] consider a serial structure and formulate an LP–model on the basis of a network representation. Setups are not part of their focus which is the reason for having continuous variables only (such approaches are used for master production scheduling). Furthermore, capacity constraints exclusively restrict the production amounts of the end item. Optimal solutions are computed by means of extreme flow considerations. A similar problem is concerned by Gabbay [Gab79] who has several serial structures being processed in parallel in mind. In contrast to Lambrecht and Vander Eecken he assumes capacity limits on all levels. Zahorik et al. [ZaThTr84] generalize Gabbay's work, but still assume serial structures. Assembly structures are investigated by Afentakis [Afe85] who transforms the problem into a job shop scheduling problem.

A first MIP–model is formulated by Haehling von Lanzenauer [HvL70a]. Remarkably to note that he considers lot sizing and scheduling simultaneously. Another early multi–level lot sizing model is that of Elmaghraby and Ginsberg [ElGi64]. An LP–model for master production scheduling in a multi–level system is given in [HvL70b].

Gorenstein [Gor70] provides a case description for planning tire production with three levels. He formulates a MIP–model, but does not present any methods. In a short note he points out that LP–based rounding methods could work fine. Vickery and Markland [ViMa86] report about experiences in using MIP–model formulations being solved with standard solvers in a real–world situation. Their field of application is a pharmaceutical company.

Zäpfel and Attmann [ZäAt80] present some kind of a tutorial for multi–level lot sizing. They assume several serial gozinto–structures being processed in parallel. A MIP–model is given on the basis of which a fixed charge problem reformulation and examples are discussed.

Ramsey and Rardin [RaRa83] consider serial structures and propose several heuristics. The basic assumption in this work is that items do not share common resources. Two of the heuristics are LP–based methods where a particular choice of the arc weights in a network flow representation makes the solution of the LP–network–problem become a feasible solution. One out of the two LP–based procedures uses the optimal solution of the uncapacitated problem [Zan69] to set the arc weights. Two other heuristics are ad hoc approaches. One makes the multi–level problem collapse into a single–level problem by making additional assumptions. The other one is a greedy heuristic. A computational study shows that the greedy heuristic gives the best results, although the idea of collapsing the gozinto–structure leads to much shorter execution times. Both LP–based approaches are worse with respect to the run–time as well as with respect to the deviation from a lower bound.

Biggs [Big85] informally discusses the problem of scheduling component parts in order to meet the demand for the end items. In a computational study he uses, among others, the total number of setups to evaluate several rules. As a result, he states that "...the use of the various rules did cause the system to respond differently...".

Harl and Ritzman [HaRi85] use modifications of uncapacitated, single–level heuristics to develop a capacity–sensitive multi–level procedure.

Billington et al. [Bil83, BiMClTh86] assume a general gozinto–structure. Some (potentially all) items share a single common resource with scarce capacity. Beside a MIP–formulation they also present a heuristic to solve the problem. The basic working principle of the heuristic is a B&B–strategy. The B&B–method enumerates over the setup variables. At each node lower bounds are computed by relaxing capacity and multi–level constraints. The uncapacitated, single–level subproblem is then heuristically solved being embedded into a Lagrangean–like iteration.

Hechtfischer [Hec91] heuristically solves problems with an assembly structure. Similar to Billington et al. he assumes that (some) items share a single common resource. But additionally, all items which require the resource must have the same low level code. Basically, his approach follows traditional MRP II ideas and operates level–by–level. If production amounts exceed the available capacity, some items are shifted into earlier periods. Guided with a Lagrangean–like penalty expression, the whole procedure iterates until a maximum number of iterations is performed.

Salomon [Sal91] introduces a simulated annealing and a tabu search heuristic for the problem earlier defined by Billington et al. [BiMClTh86]. Furthermore, he discusses LP–based heuristics for assembly structures. Among traditional rounding approaches, he also presents combinations of simulated annealing and tabu search, respectively, with LP–based rounding. Reprints of parts of this outlet are [KuSaWaMa93] and [SaKuWa93].

Roll and Karni [RoKa91] attack the multi–level lot sizing problem with a single resource as well. In contrast to Billington et al. they do not consider setup times. Starting with a first feasible solution, they pass multiple phases making changes to intermediate solutions to gain improvements.

Toklu and Wilson [ToWi92] assume assembly gozinto–structures. Furthermore, only end items require a single commonly used scarce resource. They develop a heuristic for this type of problem.

Maes et al. [Mae87, MaMClWa91, MaWa91] suggest LP–based rounding heuristics for assembly type problems based on the solution of the LP–relaxation of a reformulation as a plant location model [Ros86]. Their ideas cover simple single–pass heuristics fixing binary variables to 1, or fixing them to 1 or to 0, respectively, meanwhile solving a new LP–relaxation in–between. Another suggestion is that of curtailed B&B–procedures.

Mathes [Mat93] extends the MIP–model given by Afentakis et al. [AfGaKa84] and introduces capacity constraints whereby items do not share common resources (so–called dedicated machines). Several valid constraints are derived to improve the run–time performance of standard solvers.

Stadtler [Sta94, Sta95] compares the impact of different MIP–model formulations on the performance of standard solvers. His assumptions are quite unrestrictive: general gozinto–structures, multiple resources which are shared in common, and positive setup times. Due to his assumption that an unlimited amount of overtime per period is allowed, there exists no feasibility problem which would make the development of heuristics really hard.

Helber [Hel94, Hel95] develops heuristics for the multi–level lot sizing problem. He assumes general gozinto–structures. Items may share common resources, but no item requires more than one resource. Positive setup times are allowed. Among the proposed procedures are simulated annealing, tabu search, and genetic algorithms which guide a search in the space of setup patterns on the basis of which production plans are derived. Tempelmeier and Helber [TeHe94] test a modification of the Dixon–Silver heuristic [DiSi81]. Derstroff and Tempelmeier [Der95, TeDe93, TeDe95] use a Lagrangean relaxation of capacity constraints and inventory balances to provide a lower bound. Within each iteration an upper bound can be computed by solving an uncapacitated, multi–level problem on a level–by–level basis using the objective function including the penalty expressions for evaluation. A postprocessing stage is used to smooth capacity violations.

Clark and Armentano [ClAr95] present a heuristic for general gozinto–structures which operates in two phases. First, the Wagner–Whitin procedure is used to solve the uncapacitated problem on a level–by–level basis. Second, production in overloaded periods is shifted into earlier periods to respect the capacity constraints. Interesting to note is that items may require more than just one resource.

Following the successive planning idea, Sum and Hill [SuHi93] perform lot sizing and scheduling. They discriminate orders and operations where each order is fulfilled by doing a plenty of operations. They have both, a network (i.e. gozinto–structure in our terminology) of orders and for each order a subnetwork of operations. Operations are scheduled with a modified version of a resource constrained project scheduling algorithm [DaPa75, Pat73]. After this is done, they merge and/or break some orders which results in a new network and the whole process repeats until the recent solution is not improved. The key element of lot sizing is the merging and breaking part in this approach. The decision what to merge or to break, respectively, relies on a simple priority rule.

Dauzère–Péres and Lasserre [DaLa94a, DaLa94b, Las92] take an integrated approach in lot sizing and scheduling into account. In a first step they make lot sizing decision regarding capacity constraints. Then, they determine a sequence given fixed lot sizes. Afterwards, a new lot sizing problem is solved with new precedence constraints among items which share the same resource. These precedence relations stem from the sequence decisions just made. The whole procedure is repeated until a stopping criterion is met. For solving the lot sizing and the scheduling subproblems, respectively, they employ methods from the literature.

Haase [Haa94] gives a MIP–model for multi–level lot sizing and scheduling with a single bottleneck facility. Solution methods are, however, not provided.

Brüggemann and Jahnke [BrJa94] employ simulated annealing to find suboptimal solutions for lot sizing and scheduling problems with batch production. However, their heuristic works only for two–level gozinto–structures. The underlying idea is to proceed level by level again.

Jordan [Jor95] considers batching and scheduling. Due to restrictive assumptions, he is able to compute optimum results for small instances.

2.5 Intermediate Results

Under unconstrained capacities most authors favorite level–by–level approaches which are easy to implement and which can make use of many single–level heuristics as well as efficient exact methods. However, recent research indicates that other ideas are superior [Rou93, SiEr94b]. Due to its complexity [ArJoRo89], optimal solution procedures are unlikely to be able to solve medium– to large–sized problem instances. Hence, heuristics will dominate for practical purposes. Definitely, this holds for problems under capacity constraints.

Reviewing the work on capacitated, multi–level lot sizing reveals that most researchers consider very restrictive cases. Most of them assume a single resource only. Also, a lot of work does not deal with general gozinto–structures. Exceptions can be found in very recent outlets only [Der95, Hel94, Hel95, TeDe93, TeDe95, TeHe94]. Sophisticated heuristics for problems where items do require more than just one resource do not exist.

In summary, we find that work on multi–level lot sizing and scheduling with dynamic demand and capacity constraints indeed makes a contribution. Although its importance was recognized rather early [HvL70a], research in this field has just begun under a restrictive two–level assumption [BrJa94]. A first step away from pure successive MRP II concepts towards taking interaction effects of lot sizing and scheduling into account is done with iterative procedures [DaLa94a, DaLa94b, Las92, SuHi93].

As a consequence, all subsequent chapters shall deal with multi–level lot sizing and scheduling where the emphasis is on "and", meaning that lot sizing and scheduling decisions are done concurrently. The underlying assumptions will be far beyond what is established: general gozinto–structures with any number of levels and several scarce capacities are worth to be mentioned here. Further assumptions will be given in subsequent chapters.

Chapter 3

Problem Specifications

Until now, we have gained our understanding from informal statements only. To give a more precise definition of what is the focus of our attention, this chapter contains MIP–model formulations. In Section 3.1 we concisely describe the underlying model assumptions. Then, in Section 3.2 a MIP–model for lot sizing and scheduling with multiple machines is presented. Since the bulk of this book is about certain aspects related to this model, Section 3.3 gives some more insight and shows that the problem is indeed a non–trivial one. Sections 3.4, 3.5, and 3.6 introduce parallel machines, multiple resources, and partially renewable resources, respectively, and contain MIP–models. Remembering the literature review it becomes evident that today there are no multi–level lot sizing models and methods that rely on such general assumptions as we do below.

3.1 Basic Assumptions

Several items are to be produced in order to meet some known (or estimated) dynamic demand without backlogs and stockouts. Precedence relations among these items define an acyclic gozinto–structure of the general type. In contrast to many authors who allow demand for end items only, now, demand may occur for all items including component parts. The finite planning horizon is subdivided into a number of discrete time periods. Positive lead times are given due to technological restrictions such as cooling or transportation for instance. Furthermore, items share common resources. Some (maybe all) of them are scarce. The capacities may vary over time. Producing one item requires an item–specific amount of the available capacity. All data are assumed to be deterministic.

Items which are produced in a period to meet some future demand must be stored in inventory and thus cause item–specific holding costs. Most authors assume that the holding costs for an item must be greater than or equal to the sum of the holding costs for all immediate predecessors. They argue that holding costs are mainly opportunity costs for capital which occurs no matter a component part is assembled or not. Two reasons persuade us to make no particular assumptions for holding costs. First, as it is usual in the chemical industry for instance, keeping some component parts in storage may require ongoing additional effort such as cooling, heating, or shaking. While these parts need no special treatment when processed, storing component parts might be more expensive than storing assembled items. Second, operations such as cutting tin mats for instance make parts smaller and often easier to handle. The remaining "waste" can often be sold as raw material for other manufacturing processes. Hence, opportunity costs may decrease when component parts are assembled. However, it should be made clear that the assumption of general holding costs is the most unrestrictive one. All models and methods developed under this assumption work for more restrictive cases as well.

Each item requires at least one resource for which a setup state has to be taken into account. Production can only take place if a proper state is set up. Setting a resource up for producing a particular item incurs item–specific setup costs which are assumed to be sequence independent. Setup times are not considered. Once a certain setup action is performed, the setup state is kept up until another setup changes the current state. Hence, same items which are produced having some idle time in–between do not enforce more than one setup action. To get things straight, note that some authors use the word changeover instead of setup in this context.

The most fundamental assumption here is that for each resource at most one setup may occur within one period. Hence, at most two items sharing a common resource for which a setup state exists may be produced per period. Due to this assumption, the problem is known as the proportional lot sizing and scheduling problem (PLSP) [DrHa95, Haa94]. By choosing the length of each time period appropriately small, the PLSP is a good approximation to a continuous time axis. It refines the well–known discrete lot sizing and scheduling problem (DLSP) [Din64, Fle90, HoKo94, LaTe71, SaKrKuWa91] as well as the continuous setup lot sizing problem (CSLP) [BiMa86, KaSc85, KaKeKe87]. Both assume that at most one item may be produced per period. All three models could be classified as small bucket models since only a few (one or two) items are produced per period. In contrast to this, the well–known capacitated lot sizing problem (CLSP) [DiBaKaZi92a, EpMa87, Gün87, Hin96, KiKö94, LoCh91, MaWa88] represents a large bucket model since many items can be produced per period. Remember, the CLSP does not include sequence decisions and is thus a much "easier" problem. An extension of the single–level CLSP with partial sequence decisions can be found in [Haa93]. In [HaKi96] a large bucket single–level lot sizing and scheduling model is discussed.

3.2 Multiple Machines

An important variant of the PLSP is the one with multiple machines (PLSP–MM). Several resources (machines) are available and each item is produced on an item–specific machine. This is to say that there is an unambiguous mapping from items to machines. Of course, some items may share a common machine. Special cases are the single–machine problem for which models and methods are given in [Kim96, Kim93, Kim94a, Kim94c], and the problem with dedicated machines where items do not share a common machine. For the latter optimal solutions can be easily computed with a lot–for–lot policy [Kim94b].

Let us first introduce some notation. In Table 3.1 the decision variables are defined. Likewise, the parameters are explained in Table 3.2. Using this notation, we are now able to present a MIP–model formulation.

Symbol	Definition
I_{jt}	Inventory for item j at the end of period t.
q_{jt}	Production quantity for item j in period t.
x_{jt}	Binary variable which indicates whether a setup for item j occurs in period t ($x_{jt} = 1$) or not ($x_{jt} = 0$).
y_{jt}	Binary variable which indicates whether machine m_j is set up for item j at the end of period t ($y_{jt} = 1$) or not ($y_{jt} = 0$).

Table 3.1: Decision Variables for the PLSP–MM

$$\min \sum_{j=1}^{J} \sum_{t=1}^{T} (s_j x_{jt} + h_j I_{jt}) \tag{3.1}$$

subject to

$$I_{jt} = I_{j(t-1)} + q_{jt} - d_{jt} - \sum_{i \in \mathcal{S}_j} a_{ji} q_{it} \qquad \begin{array}{l} j = 1, \ldots, J \\ t = 1, \ldots, T \end{array} \tag{3.2}$$

$$I_{jt} \geq \sum_{i \in \mathcal{S}_j} \sum_{\tau = t+1}^{\min\{t+v_j, T\}} a_{ji} q_{i\tau} \qquad \begin{array}{l} j = 1, \ldots, J \\ t = 0, \ldots, T-1 \end{array} \tag{3.3}$$

$$\sum_{j \in \mathcal{J}_m} y_{jt} \leq 1 \qquad \begin{array}{l} m = 1, \ldots, M \\ t = 1, \ldots, T \end{array} \tag{3.4}$$

$$x_{jt} \geq y_{jt} - y_{j(t-1)} \qquad \begin{array}{l} j = 1, \ldots, J \\ t = 1, \ldots, T \end{array} \tag{3.5}$$

$$p_j q_{jt} \leq C_{m_j t} (y_{j(t-1)} + y_{jt}) \qquad \begin{array}{l} j = 1, \ldots, J \\ t = 1, \ldots, T \end{array} \tag{3.6}$$

$$\sum_{j \in \mathcal{J}_m} p_j q_{jt} \leq C_{mt} \qquad \begin{array}{l} m = 1, \ldots, M \\ t = 1, \ldots, T \end{array} \tag{3.7}$$

Symbol	Definition
a_{ji}	"Gozinto"–factor. Its value is zero if item i is not an immediate successor of item j. Otherwise, it is the quantity of item j that is directly needed to produce one item i.
C_{mt}	Available capacity of machine m in period t.
d_{jt}	External demand for item j in period t.
h_j	Non–negative holding cost for having one unit of item j one period in inventory.
I_{j0}	Initial inventory for item j.
$\mathcal{J}_m$	Set of all items that share the machine m, i.e. $\mathcal{J}_m \stackrel{def}{=} \{j \in \{1,\ldots,J\} \mid m_j = m\}$.
J	Number of items.
M	Number of machines.
m_j	Machine on which item j is produced.
p_j	Capacity needs for producing one unit of item j.
s_j	Non–negative setup cost for item j.
$\mathcal{S}_j$	Set of immediate successors of item j, i.e. $\mathcal{S}_j \stackrel{def}{=} \{i \in \{1,\ldots,J\} \mid a_{ji} > 0\}$.
T	Number of periods.
v_j	Positive and integral lead time of item j.
y_{j0}	Unique initial setup state.

Table 3.2: Parameters for the PLSP–MM

$$y_{jt} \in \{0,1\} \qquad \begin{array}{l} j = 1,\ldots,J \\ t = 1,\ldots,T \end{array} \tag{3.8}$$

$$I_{jt}, q_{jt}, x_{jt} \geq 0 \qquad \begin{array}{l} j = 1,\ldots,J \\ t = 1,\ldots,T \end{array} \tag{3.9}$$

The objective (3.1) is to minimize the sum of setup and holding costs. Equations (3.2) are the inventory balances. At the end of a period t we have in inventory what was in there at the end of period $t-1$ plus what is produced minus external and internal demand. To fulfill internal demand we must respect positive lead times. Restrictions (3.3) guarantee so. Constraints (3.4) make sure that the setup state of each machine is uniquely defined at the end of each period. Those periods in which a setup happens are spotted by (3.5). Note that idle periods may occur in order to save setup costs. Due to (3.6) production can only take place if there is a proper setup state either at the beginning or at the end of a particular period. Hence, at most two items can be manufactured on each machine per period. Capacity constraints are formulated in (3.7). Since the right hand side is a constant, overtime is not available. (3.8) define the binary–valued setup state variables, while (3.9) are simple non–negativity conditions. The reader may convince himself that due to (3.5) in combination with (3.1) setup variables x_{jt} are indeed zero–one valued. Hence, non–negativity conditions are sufficient for these. For letting inventory variables I_{jt} be non–negative backlogging cannot occur.

A DLSP–like model can be derived from the PLSP–model by adding

$$p_j q_{jt} = C_{m_j t} y_{jt} \qquad \begin{array}{l} j = 1, \ldots, J \\ t = 1, \ldots, T \end{array} \qquad (3.10)$$

to the set of constraints. It can easily be verified that now at most one item can be produced per period. Typical for the DLSP is the so–called "all–or–nothing" production. Having these equations introduced, the resulting model can be simplified. (3.6) and (3.7) are superfluous and can be dropped. Moreover, variables q_{jt} can be eliminated by substitution. Since adding constraints does not decrease the optimal objective function value, we have given a proof that the optimal objective function value of a PLSP–instance is less than or equal to the optimal objective function value of the corresponding DLSP–instance. Moreover, if lots of the same item are produced in different periods, then idle periods cannot occur in–between without enforcing a new setup which is in contrast to the PLSP. Furthermore, whereas in the PLSP–model the unequals sign in (3.4) can be replaced with an equals sign without changing anything, the DLSP–model must have an unequals sign. It should be remarked, that traditional DLSP–model formulations do not consider time varying capacities, i.e. $C_{m1} = \ldots = C_{mt} = \ldots = C_{mT}$ for all $m = 1, \ldots, M$.

A CSLP–model formulation can be derived from the PLSP–model be adding

$$p_j q_{jt} \leq C_{m_j t} y_{jt} \qquad \begin{array}{l} j = 1, \ldots, J \\ t = 1, \ldots, T \end{array} \tag{3.11}$$

to the set of constraints. The fact that at most one item can be produced per period should again be clear when looking at (3.4). Similar to what was done when we derived the DLSP–model, some steps of simplification can reduce the size of the resulting model. As in the DLSP–case above, (3.6) and (3.7) can be dropped. Again, we now have a formal proof that the optimal objective function value of a PLSP–instance is always less than or equal to the objective function value of the corresponding CSLP–instance. In analogy to the PLSP, the CSLP also allows idle periods in order to save setup costs.

In summary, the PLSP can be seen as a generalization of the DLSP and the CSLP. With minor modifications all methods developed for the PLSP should be applicable to the latter ones as well. However, the reverse is not true. Efficient methods making use of restrictive assumptions need not work well for the PLSP. It comes out that an optimal solution for a PLSP instance is less than or equal to an optimal solution of the DLSP or the CSLP using the same data set. Since heuristic methods for the multi–level DLSP or the multi–level CSLP do not exist, there is no such benchmark for suboptimal procedures available for the PLSP.

To transform the PLSP–model into a CLSP–model we must drop (3.4). The constraints

$$p_j q_{jt} \leq C_{m_j t} y_{jt} \qquad \begin{array}{l} j = 1, \ldots, J \\ t = 1, \ldots, T \end{array} \tag{3.12}$$

and

$$x_{jt} = y_{jt} \qquad \begin{array}{l} j = 1, \ldots, J \\ t = 1, \ldots, T \end{array} \tag{3.13}$$

are introduced instead. Note that setup costs are charged now in every period in which production takes place. Again, the resulting model can be drastically reduced. (3.5) and (3.6) are redundant. Moreover, we can eliminate the variables x_{jt} by substitution.

Usually, $v_j = 0$ is assumed for all or some $j = 1, \ldots, J$ since we have large (time) buckets in mind. This makes (3.3) obsolete for these items. It should be clear that once we give up (3.4) the resulting model does not support sequence decisions any more. A general statement about the relation between optimal objective function values for PLSP–instances and corresponding CLSP–instances cannot be made.

3.3 Problem Insights

3.3.1 Complexity Considerations

Only very little theoretical research has been done on complexity issues[1] for lot sizing (and scheduling) (see [PoWa92] for a review). Optimizing the single–level CLSP is known to be NP–hard [BiYa82]. So must be optimizing the multi–level CLSP. The complexity of the single–level, single–machine DLSP is examined in [Sal91, SaKrKuWa91]. While the feasibility problem is polynomially solvable, the optimization problem is claimed to be NP–hard. In [Brü95] it is pointed out that the presented proof of this claim is false. Nevertheless, a correct proof also given in [Brü95] shows that the optimization problem is (strongly) NP–hard. So must be the multi–level DLSP.

Theoretical results for any variant of the CSLP or the PLSP are not published yet. But, as we have learned from experience, for most multi–level PLSP instances it is quite a formidable task to find even a feasible solution. Using heuristics to attack the PLSP–MM (and its extensions) thus seems to be an extremely good piece of advice.

3.3.2 Derived Parameters

Throughout this book it will be helpful to have some notation for certain information that can be derived from the parameters

[1] An introduction into the theory of complexity would need more space than this subsection and is out of the scope of this book. We refer to [GaJo79].

defined above. Also, evaluating the parameters for some values not directly given deepens the understanding which is a motivation on its own right.

To start with, let us analyze the gozinto–structure a bit more. We already have defined the set of immediate successors $\mathcal{S}_j$ of an item j. The set of immediate predecessors $\mathcal{P}_j$ of an item j can be defined as follows:

$$\mathcal{P}_j \stackrel{def}{=} \{i = 1, \ldots, J \mid j \in \mathcal{S}_i\} \qquad j = 1, \ldots, J \tag{3.14}$$

On the basis of this, the set of all successors $\bar{\mathcal{S}}_j$ and the set of all predecessors $\bar{\mathcal{P}}_j$, respectively, of an item j are given as

$$\bar{\mathcal{S}}_j \stackrel{def}{=} \mathcal{S}_j \cup \left\{k \in \bar{\mathcal{S}}_i \mid i \in \mathcal{S}_j\right\} \qquad j = 1, \ldots, J \tag{3.15}$$

and

$$\bar{\mathcal{P}}_j \stackrel{def}{=} \mathcal{P}_j \cup \left\{k \in \bar{\mathcal{P}}_i \mid i \in \mathcal{P}_j\right\} \qquad j = 1, \ldots, J. \tag{3.16}$$

Producing one unit of item j triggers the production of preceding items. The internal demand id_{ji} for an item i caused by producing one item j is computed as

$$id_{ji} \stackrel{def}{=} \begin{cases} 1 & \text{, if } j = i \\ 0 & \text{, if } i \notin \bar{\mathcal{P}}_j \\ \sum\limits_{k \in \mathcal{P}_j} a_{kj} id_{ki} \left(= \sum_{k \in \mathcal{S}_i} a_{ik} id_{jk}\right) & \text{, otherwise} \end{cases} \tag{3.17}$$

for $j, i = 1, \ldots, J$ where it should be emphasized that the case $j = i$ does not indicate cyclic structures, but eases the notation.

The expression

$$dep_j \stackrel{def}{=} \begin{cases} 0 & \text{, if } \mathcal{P}_j = \emptyset \\ \max\limits_{i \in \mathcal{P}_j} \{v_i + dep_i\} & \text{, otherwise} \end{cases} \qquad j = 1, \ldots, J \tag{3.18}$$

defines the depth of an item. The low level code llc_j of an item j, informally given in Chapter 2, can be restated more formally:

$$llc_j \stackrel{def}{=} \begin{cases} 0 & \text{, if } \mathcal{S}_j = \emptyset \\ 1 + \max\limits_{i \in \mathcal{S}_j} \{llc_i\} & \text{, otherwise} \end{cases} \qquad j = 1, \ldots, J \tag{3.19}$$

The net requirement nr_j of an item j can be computed in an item–by–item stepwise manner. To describe the calculation we assume a technological ordering without loss of generality, i.e.

$$j < i \Rightarrow j \notin \bar{\mathcal{P}}_i \qquad j, i = 1, \ldots, J$$

must hold. By the way, note

$$j < i \Rightarrow llc_j \leq llc_i \qquad j, i = 1, \ldots, J$$

would be fine as well, but gives less freedom for labeling items. Remember that only acyclic gozinto–structures are taken into account. The adjacency matrix representation of a gozinto–structure can thus always be written in triangular form

$$(a_{ji})_{j,i=1,\ldots,J} = \begin{pmatrix} 0 & \cdots & \cdots & \cdots & 0 \\ a_{21} & \ddots & & & \vdots \\ a_{31} & a_{32} & \ddots & & \vdots \\ \vdots & & & \ddots & \vdots \\ a_{J1} & a_{J2} & \cdots & a_{J(J-1)} & 0 \end{pmatrix}.$$

Furthermore, we make use of an auxiliary function

$$\chi_1(x) \stackrel{def}{=} \frac{x + \mid x \mid}{2} = \begin{cases} x & , \text{if } x \geq 0 \\ 0 & , \text{otherwise} \end{cases}. \tag{3.20}$$

The procedure to determine nr_j can now be given as in Table 3.3. Without loss of generality, we subsequently assume that every item has a positive net requirement.

$nr_j := \sum_{t=1}^{T} d_{jt}$ for all $j = 1, \ldots, J$.
for $j = 1$ to $j = J$

$$nr_j := \chi_1\left(nr_j + \sum_{i \in \mathcal{S}_j} a_{ji} nr_i - I_{j0}\right).$$

Table 3.3: A Method to Compute the Net Requirement

3.3.3 Another Point of View: Gozinto–Trees

For illustrating graphically which item causes what internal demand we will convert matrices of gozinto–factors into acyclic but general graphs throughout the book. While these pictures are compact and easy to read, they misleadingly reflect the precedence relations for scheduling items.[2] A representation that directly reveals the precedence relations in a feasible schedule is gained by converting a general gozinto–graph into an assembly structure by copying nodes with more than one successor. Figure 3.1 gives an example (see also [AfGa86] where this point of view helps to solve problems with complex product structures). Note, both representations contain exactly the same information and thus are equivalent. We will use the term gozinto–tree to refer to the gozinto–structure when converted into an assembly structure. It should be clear that the gozinto–tree actually is a forest, if the gozinto–structure contains more than one end item.

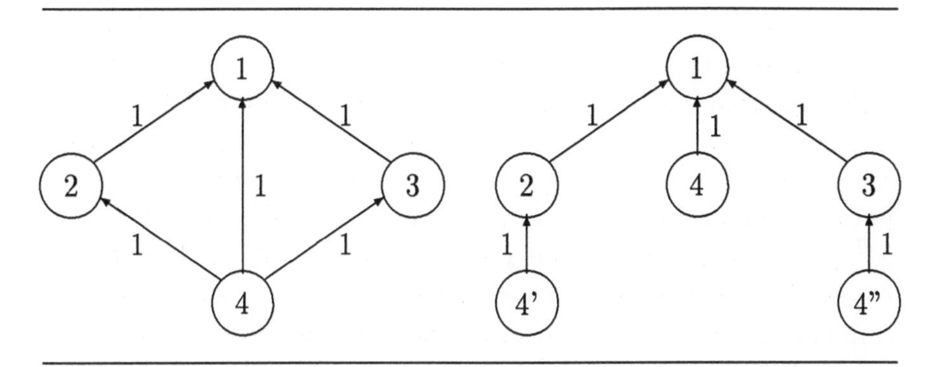

Figure 3.1: Converting a Gozinto–Structure into a Gozinto–Tree

Since gozinto–trees will play a key role in the methods to be presented, we give here a formal definition of a gozinto–tree. As-

[2]Compare the graphical representation of a gozinto–structure with an activity–on–node network as used in project scheduling. In the latter one, precedence relations in the graph correspond to precedence relations that must be respected in every feasible schedule.

sume a matrix of gozinto–factors to be given. Then, a set denoted as Γ_j represents the gozinto–tree with item j being its root node. This set is recursively defined by

$$\Gamma_j \stackrel{def}{=} \{(j, \bigcup_{i \in \mathcal{P}_j} \Gamma_i)\} \qquad j = 1, \ldots, J. \tag{3.21}$$

The pair in the set Γ_j represents the root node in the corresponding gozinto–(sub–)tree. At the first position, we have the number of the item. At position number two, all immediately preceding nodes are shown. E.g., consider the gozinto–tree given in Figure 3.1:

$$\Gamma_1 = \{(1, \{(2, \{(4, \emptyset)\}), (3, \{(4, \emptyset)\}), (4, \emptyset)\})\}$$

3.3.4 Initial Inventory

In the single–level case one can assume without loss of generality that the initial inventory I_{j0} is zero for all items $j = 1, \ldots, J$. The reason for this is that positive stock levels can be mapped to external demands before the planning process begins. The rule to do so is very simple. Initial inventory of an item j is used to fulfill external demand in a period t only if all external demands in periods τ where $1 \leq \tau < t$ are also met by the initial inventory. More formally, $I_{j0} > \sum_{\tau=1}^{t-1} d_{j\tau}$ must hold if external demand in period t should be met by initial inventory. In the case that $I_{j0} \geq \sum_{t=1}^{T} d_{jt}$ holds, item j need not be produced at all. It is trivial to see that the preprocessed instance is feasible if and only if the original instance is feasible. And, the optimum objective function value of the preprocessed instance equals the optimum objective function value of the original instance.

In the multi–level case there are some strings attached which do not keep things that easy. A small example certainly brings this out best. Suppose $J = 2$ items are manufactured on a single machine. The gozinto–structure is given in Figure 3.2. Let the planning horizon be $T = 4$. All relevant data are given in Table 3.4. Points worth to be highlighted are that the machine is initially set up for item 1 ($y_{110} = 1$) and that there is a positive initial inventory for item 2 ($I_{20} = 10$).

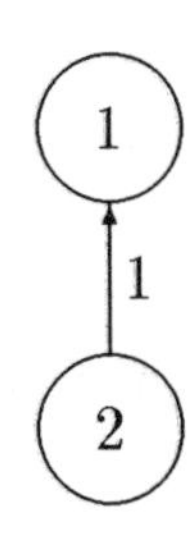

Figure 3.2: A Gozinto–Structure with Two Items

d_{jt}	$t=1$	2	3	4	h_j	s_j	p_j	v_j	y_{j0}	I_{j0}
$j=1$				10	20	900	1	1	1	
$j=2$		10			10	800	1	1		10
C_{1t}	15	15	15	15						

Table 3.4: Parameters of the Example

Now assume that we proceed as we would do in the single–level case. That is, before we start with the planning process we eliminate the initial inventory during a preprocessing phase. Since the initial inventory meets the external demand for item 2 in period 2, the result is an instance without initial inventory where external demand for item 1 is left only. The optimum solution for the remaining instance is defined by a lot–for–lot policy ($q_{23} = q_{14} = d_{14} = 10$). The optimum objective function value is $s_1 + s_2 + h_2 I_{20} + h_2 a_{21} q_{14} = 1,900$.

The optimum solution for the unmodified instance is given in Table 3.5. Its objective function value is $s_2 + h_2(I_{20} - a_{21}q_{11}) + h_1(T-1)q_{11} + h_1(T-2)q_{12} = 1,350 < 1,900$.

Apparently, something goes wrong when eliminating initial inventory to get an instance with zero initial inventory. As pointed out for the single–level case, initial inventory of an item j can be mapped to demand in a period t without harm only if all other

q_{jt}	$t=1$	2	3	4
$j=1$	5	5		
$j=2$		10		
j with $y_{jt}=1$	1	2	2	2

Table 3.5: An Optimum Solution of the Example

demand for item j in earlier periods can also be met by the initial inventory. In the multi–level case we have external as well as internal demand. But only the external demand is known by time. The time when internal demand occurs is not known unless a production plan is constructed. Hence, in general there cannot be a preprocessing procedure that eliminates initial inventory. As a consequence, multi–level lot sizing and scheduling procedures must take initial inventory into account.

Eventually, a rather unexpected phenomenon should be mentioned which may occur if positive initial inventory comes along with multi–level gozinto–structures. Surprisingly, initial inventory may be the reason for producing items without any demand for it. Consider, for instance, the example above, this time, assuming $d_{jt} = 0$ for $j = 1, \ldots, J$ and $t = 1, \ldots, T$. In an optimum solution no production takes place. The optimum objective function value, however, is not zero, but $h_2 I_{20} T = 400$ for keeping 10 units of item 2 four periods in inventory. Suppose, $h_1 < h_2$. Then, the optimum result would be to produce $q_{11} = 10$ units of item 1 in period 1 which uses the initial inventory of item 2 up. Since there exists no demand for item 1, we have $I_{1T} = 10$, and the optimum objective function value is $h_1 q_{11} T < h_2 I_{20} T$.

3.3.5 Schedules

Discovering certain properties that optimum solutions must have allows to reduce the set of feasible solutions among which an optimum solution is to be and among which we should try to find a good feasible solution. We will now show that due to the combination of multi–level gozinto–structures and general holding costs a "nice" property of optimum schedules is no longer valid, though it is in the single–level case.

To avoid unnecessary repetitions let us assume once and for all the following parameters in this subsection: $M = 1$, $T = 10$, $C_{1t} = 15$, $p_j = 1$, $I_{j0} = 0$, and $v_j = 1$ for all $j = 1, \ldots, J$ and $t = 1, \ldots, T$. J will vary. The parameters y_{jm_j0} and s_j are of no importance here and can be chosen arbitrarily.

For the first example which reveals a property of optimum schedules in the single–level case, let $J = 3$. External demand is given in Table 3.6. The holding cost $h_j > 0$ can be chosen arbitrarily for all $j = 1, \ldots, J$.

d_{jt}	$t = 1$	2	3	4	5	6	7	8	9	10
$j = 1$					20					20
$j = 2$		20								
$j = 3$									20	

Table 3.6: External Demand

Figure 3.3–(a) shows a feasible schedule. Note, the lot for item 1 fulfills the external demand in period 5 as well as the external demand in period 10. Suppose, that the sequence of lots should not be changed and wonder if there may exist another schedule (with the same sequence of lots and therefore the same total setup costs) with a lower objective function value (which means with lower total holding costs). The answer is yes, and the schedule with lowest objective function value (while respecting the sequence of lots) is depicted in Figure 3.3–(b). It is essential to understand that idle periods do not enforce additional setups.

The difference between both schedules is that in case (b) all items are shifted right–most without changing the sequence and without violating due dates. In general, given a sequence of lots to be produced on a common machine, the (unique) schedule with the lowest objective function value is the one which schedules all items right–most. Adopting the terminology of scheduling theory

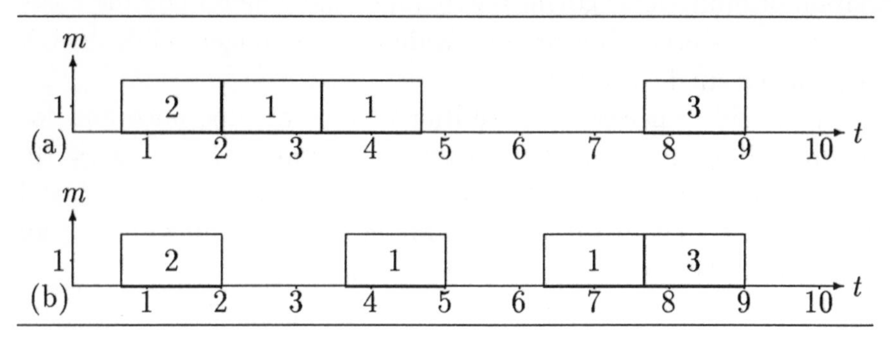

Figure 3.3: Two Schedules with the Same Sequence of Lots

[SpKoDr95] here, such schedules are called semi–active schedules.[3] As we will now show, the property that an optimum schedule must be a semi–active one is valid for single–level structures only unless additional (restrictive) assumptions are made.

So, let us have a look at another example with $J = 2$ and a two–level gozinto–structure (see Figure 3.2). External demand is given in Table 3.7. Suppose here that $h_2 > h_1$ holds.

d_{jt}	$t = 1$	2	3	4	5	6	7	8	9	10
$j = 1$					20					20
$j = 2$										

Table 3.7: External Demand

Figure 3.4–(a) provides a feasible schedule. For the given sequence of lots this is the unique semi–active schedule. Figure 3.4–(b), however, shows a schedule with lower objective function value respecting the sequence of lots. This one is not semi–active.

The reason why the second schedule has a lower objective function value is that holding item 2 is more expensive than holding

[3]In contrast to our interpretation of semi–active schedules as right–most schedules, in scheduling theory left–most schedules are usually meant. The reason for this is a difference in the objective function. Minimizing the makespan is, for instance, a widespread objective.

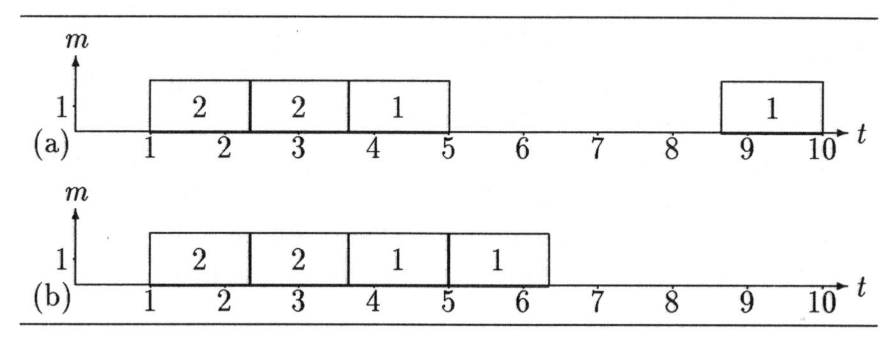

Figure 3.4: Two Schedules with the Same Sequence of Lots

item 1. All lots for item 1 should therefore be scheduled close to the lot of item 2 which meets the internal demand. In general, a sufficient condition for an optimum schedule to be semi–active is

$$h_j \geq \sum_{i \in \mathcal{P}_j} a_{ij} h_i \qquad j = 1, \ldots, J \tag{3.22}$$

which is a restrictive assumption not made here (see Section 3.1). Nevertheless, as a result we have the fact that given a certain sequence of lots, those items j which fulfill (3.22) should be scheduled right–most.

3.3.6 Lot Splitting

Looking at all examples given up to here again, one might conclude that lot sizing (and scheduling) is about grouping demands together (and finding a (sub–)optimal schedule for these lots). Thus, lot sizes appear to be the sum of order sizes. While this is true for single–level, uncapacitated problems [WaWh58], lot sizing with multi–level gozinto–structures and/or capacity constraints needs to take splitting lots — splitting order sizes would be more precise[4]

[4]Note, lot splitting in our context does not necessarily mean to divide a lot into sublots and to process each sublot in parallel on identical facilities [APICS95a]. Sublots may also be processed consecutively (with other lots in–between).

— into account and is therefore much more difficult as we will point out. The problem of lot sizing (and scheduling) where only grouping is allowed but no splitting, is called a batching (and sequencing) problem [Jor95, UnKi92, WoSp92].[5]

Let us investigate the uncapacitated, multi-level case first.[6] Assume $M = 2$ and a gozinto-structure as given in Figure 3.2. Table 3.8 gives the parameters that are of interest. Note that we have positive initial inventory. Furthermore, assume $h_2 > h_1$.

d_{jt}	$t = 1$	2	3	m_j	I_{j0}	v_j
$j = 1$			20	1		1
$j = 2$				2	10	1

Table 3.8: Parameters of an Example without Capacity Constraints

The optimum production plan is depicted in Figure 3.5. It is remarkable to see, that the external demand of item 1 is split into two lots of size 10. Together with the results in Subsection 3.3.4 we now have that if generality should not be lost, multi-level gozinto-structures imply that lot splitting must be taken into account in order to find optimum solutions.

Noteworthy to say that in capacitated cases lot splitting is not a problem introduced with multi-level structures. Suppose

[5]To convert the lot sizing problem into a batching problem we need to add additional constraints to the PLSP-MM-model formulation. If we assume $I_{j0} = 0$ for all $j = 1, \ldots, J$, the restrictions

$$I_{j(t-1)} - \sum_{i \in \mathcal{S}_j} \sum_{\tau=t}^{\min\{t+v_j-1,T\}} a_{ji} q_{i\tau} \leq B(1 - x_{jt}) \qquad \begin{array}{l} j = 1, \ldots, J \\ t = 1, \ldots, T \end{array}$$

where B is a large number would work fine. Since batching is not in our interest, we omit further comments and leave the discussion of the case of positive initial inventory out.

[6]More formally, assume the problem under concern be defined by the PLSP-MM-model without the capacity constraints (3.7). The capacity parameters in (3.6) must be replaced with a large number, and all p_j-values can be replaced by one.

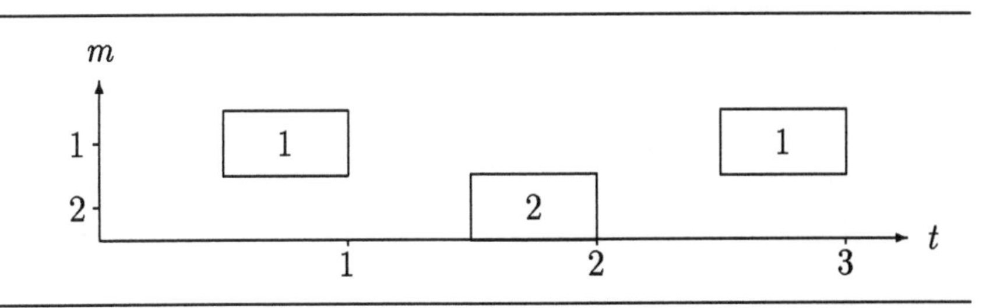

Figure 3.5: Lot Splitting in Uncapacitated Cases

a single–level problem instance where $M = 1$, $T = 3$, and all other relevant parameters are given in Table 3.9. Setup costs are of no relevance. Note that the machine is initially set up for item 2. Let holding costs be arbitrarily chosen with respect to $h_1 > h_2$. There is only one feasible solution without lot splitting (see Figure 3.6–(b)). As we can see in Figure 3.6–(a), a feasible solution with lot splitting would decrease the objective function value by $10(h_1 - h_2) > 0$.

d_{jt}	$t = 1$	2	3	p_j	y_{j0}	I_{j0}
$j = 1$		20		1		
$j = 2$			30	1	1	
C_{1t}	20	20	20			

Table 3.9: Parameters of the Example

A stronger observation is that some instances do have feasible solutions with lot splitting only. Again, suppose the parameters in Table 3.9 with slightly modified data, i.e. $d_{21} = 10$ and $y_{20} = 0$. To understand the logic of the example, it is important to be aware of the fact that the machine is not set up for any item initially. Figure 3.7 shows the unique feasible solution.

It is interesting to see that there are instances for which all feasible solutions are without lot splitting and no feasible solution with lot splitting exists. Once more, suppose the parameters given

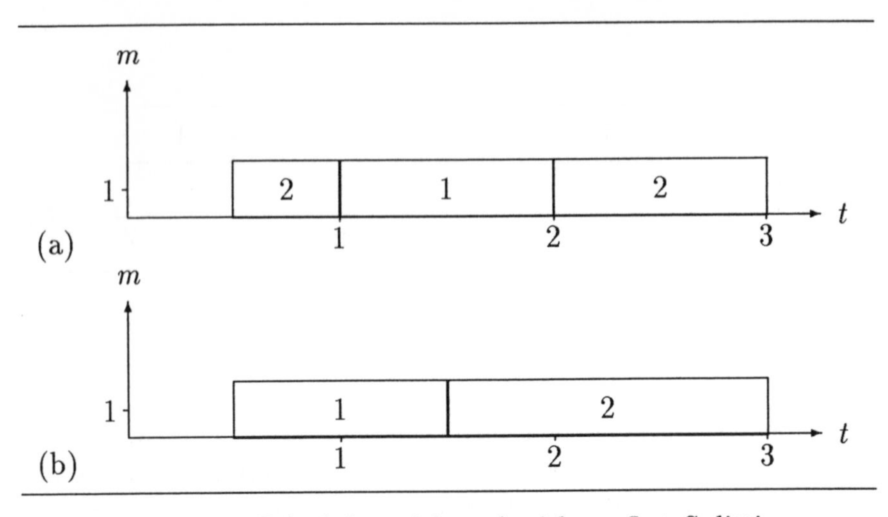

Figure 3.6: Schedules with and without Lot Splitting

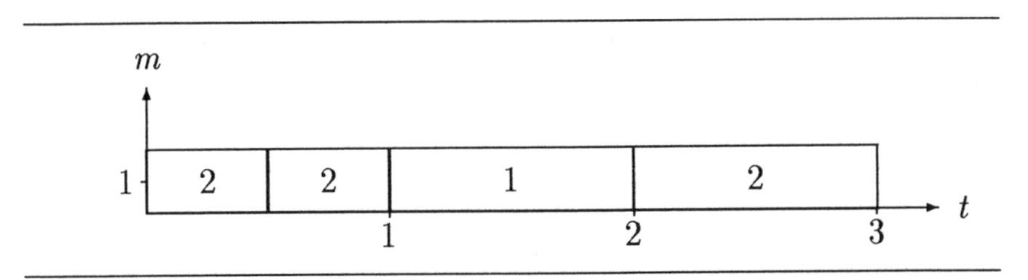

Figure 3.7: The Unique Feasible Solution of the Example

in Table 3.9 with slightly modified data, i.e. $d_{12} = 25$ and $y_{210} = 0$. Then, there is only one semi–active feasible solution which is depicted in Figure 3.8 where no lot splitting occurs.

In general, we have the following results.

- If there exists a feasible solution for an instance without lot splitting, the existence of feasible solutions with lot splitting is not guaranteed.

- If there exists a feasible solution for an instance with lot splitting, the existence of feasible solutions without lot split-

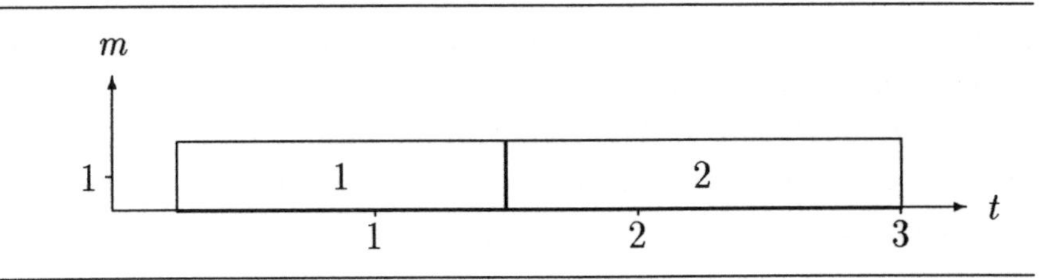

Figure 3.8: The Optimum Solution of the Example

ting is not guaranteed, too.

- If both feasible solutions with and without lot splitting exist, the optimum solution is not necessarily without lot splitting.

Instances which have both types of feasible solutions are guaranteed to have an optimum solution without lot splitting if

$$h_j p_j \leq h_i p_i \qquad j, i = 1, \ldots, J \tag{3.23}$$

holds. In other words, we would have to assume that

$$h_j p_j = h_i p_i \qquad j, i = 1, \ldots, J \tag{3.24}$$

holds, to guarantee that for any instance having a feasible solution in which no lot splitting occurs an optimum solution can be found without lot splitting. As already stated above, we do not assume this here.

3.3.7 Some Valid Constraints

In this subsection we present some additional constraints which are proven to be valid in the sense that optimum solutions fulfilling these do exist. If there are any feasible solutions of the original PLSP–MM–model[7] violating the new constraints, the valid constraints reduce the solution space and thus help to find

[7] In subsequent sections we will discuss extensions of the PLSP–MM–model. The valid constraints need to undergo a minor modification to apply there, too. However, the underlying ideas remain true.

(sub–)optimum solutions (e.g. when standard MIP–solvers are employed or heuristics are to be developped) or at least help to gain a better understanding.

Valid constraint 1: The following equations determine the inventory at the end of each period. They can be used to eliminate the inventory variables.

$$I_{jt} = I_{j0} + \sum_{\tau=1}^{t} \left(q_{j\tau} - d_{j\tau} - \sum_{i \in \mathcal{S}_j} a_{ji} q_{i\tau} \right) \qquad \begin{array}{l} j = 1, \ldots, J \\ t = 1, \ldots, T \end{array} \tag{3.25}$$

Proof: Trivial. It is a recursive evaluation of (3.2). □

Valid constraint 2: If there is no initial inventory, items must be produced only if there is a positive requirement.[8] This may fix the value of some decision variables to zero.

$$q_{jt} \leq nr_j \qquad \begin{array}{l} j = 1, \ldots, J \\ t = 1, \ldots, T \end{array} \tag{3.26}$$

$$x_{jt} \leq nr_j \qquad \begin{array}{l} j = 1, \ldots, J \\ t = 1, \ldots, T \end{array} \tag{3.27}$$

$$y_{jt} \leq nr_j \qquad \begin{array}{l} j = 1, \ldots, J \\ t = 1, \ldots, T \end{array} \tag{3.28}$$

Proof: Trivial.[9] □

Valid constraint 3: For each item $j = 1, \ldots, J$ with $nr_j > 0$ the machine m_j must be set up for this item at least once. This holds for all feasible solutions.

$$nr_j \sum_{t=0}^{T} y_{jt} \geq nr_j \qquad j = 1, \ldots, J \tag{3.29}$$

Proof: Trivial. □

Valid constraint 4: Assume that $q_{jt} > 0$ should imply $q_{jt} \geq 1$ for $j = 1, \ldots, J$ and $t = 1, \ldots, T$. A setup for an item j where $j = 1, \ldots, J$ should only take place if the internal demand is met

[8] Remember Subsection 3.3.4 where it is shown that the following unequalities are not necessarily true when we face positive initial inventory.

[9] Note, to let the constraints for setup state variable y_{jt} be valid (3.4) must not contain an equals sign.

so that at least one item can be produced. This enforces some setup variables to be zero.

$$a_{ij}x_{jt} \leq I_{i(t-1)} + I_{it} \qquad \begin{array}{l} j = 1,\ldots,J \\ i \in \mathcal{P}_j \\ t = 1,\ldots,T \end{array} \tag{3.30}$$

Proof: Let us consider items j and i and period t. If there is an optimum solution with $x_{jt} = 0$ the resulting inequality is true because of (3.9). So, assume there is an optimum solution with $x_{jt} = 1$. Due to (3.5) we have $y_{j(t-1)} = 0$ and $y_{jt} = 1$. Using (3.6) either $q_{jt} > 0$ or $q_{j(t+1)} > 0$ (if $t < T$) or both must hold, otherwise $y_{jt} = 0$ and hence $x_{jt} = 0$ would give a lower objective function value or would enforce $x_{j(t+1)} = 1$ (if $t < T$) resulting in another optimum solution. Suppose, $q_{jt} > 0$. Then we are done, since $a_{ij}x_{jt} \overset{q_{jt}>0}{\leq} a_{ij}q_{jt} \overset{(3.3)}{\leq} I_{i(t-1)} \overset{(3.9)}{\leq} I_{i(t-1)} + I_{it}$. Suppose, $t < T$ and $q_{j(t+1)} > 0$. Analogously, we are done again, because $a_{ij}x_{jt} \overset{q_{j(t+1)}>0}{\leq} a_{ij}q_{j(t+1)} \overset{(3.3)}{\leq} I_{it} \overset{(3.9)}{\leq} I_{i(t-1)} + I_{it}$. □

Valid constraint 5: Similar to (3.30) we can formulate new constraints for production quantities and setup state variables which set these decision variables to zero in some cases.

$$a_{ij}q_{jt} \leq I_{i(t-1)} + I_{it} \qquad \begin{array}{l} j = 1,\ldots,J \\ i \in \mathcal{P}_j \\ t = 1,\ldots,T \end{array} \tag{3.31}$$

$$a_{ij}(y_{jt} - y_{j(t-1)}) \leq I_{i(t-1)} + I_{it} \qquad \begin{array}{l} j = 1,\ldots,J \\ i \in \mathcal{P}_j \\ t = 1,\ldots,T \end{array} \tag{3.32}$$

Proof: Just follow the lines of (3.30). □

Valid constraint 6: A machine should switch the setup state in period $t = 1,\ldots,T$ for producing an item $j = 1,\ldots,J$ only if items $i \in \mathcal{P}_j$ with $I_{i0} = 0$ have a chance to be manufactured in advance. This condition lets some setup state variables be zero.

$$y_{jt} \leq y_{j0} + \sum_{\tau=0}^{t} y_{i\tau} \qquad \begin{array}{l} j = 1,\ldots,J \\ i \in \{k \in \mathcal{P}_j \mid I_{k0} = 0 \text{ and } m_k \neq m_j\} \\ t = 1,\ldots,T \end{array} \tag{3.33}$$

$$y_{jt} \leq \sum_{\tau=0}^{t} y_{i\tau} \qquad \begin{array}{l} j = 1,\dots,J \\ i \in \{k \in \mathcal{P}_j \mid I_{k0} = 0 \\ \qquad \text{and } m_k = m_j\} \\ t = 1,\dots,T \end{array} \tag{3.34}$$

Proof: We only prove the second inequality, because the first basically follows the same ideas. The difference is that in the first case an initial setup state can be kept up. Let the items under consideration be j and i where $i \in \{k \in \mathcal{P}_j \mid I_{k0} = 0 \text{ and } m_k = m_j\}$. Suppose that $\sum_{\tau=0}^{t} y_{i\tau} = 1$. Due to (3.8) the resulting inequality is true. So, assume there is an optimum solution with $\sum_{\tau=0}^{t} y_{i\tau} = 0$. With (3.6) we get $q_{it} = 0$ for all $t = 1,\dots,t$. Because $I_{i0} = 0$, we also have $I_{i(t-1)} = I_{it} = 0$ using (3.2). (3.3) implies $q_{jt} = q_{j(t+1)} = 0$. Thus, there is an optimum solution with $y_{jt} = 0$. □

Valid constraint 7: Similar to the setup state variables we can formulate constraints for the setup variables to enforce some of these to be zero.

$$x_{jt} \leq y_{i0} + \sum_{\tau=1}^{t} x_{i\tau} \qquad \begin{array}{l} j = 1,\dots,J \\ i \in \{k \in \mathcal{P}_j \mid I_{k0} = 0\} \\ t = 1,\dots,T \end{array} \tag{3.35}$$

Proof: Note, $y_{i0} + \sum_{\tau=1}^{t} x_{i\tau} = 0$ implies $\sum_{\tau=0}^{t} y_{i\tau} = 0$ due to (3.5) for all $i = 1,\dots,J$. From then on the proof is similar to (3.34). □

Valid constraint 8: If there exists no initial inventory and if there is no future demand for an item $j = 1,\dots,J$ there is no need to keep up the setup state or to produce anything. Again, this restricts the value of some decision variables

$$q_{jt} \leq \sum_{\tau=t}^{T} d_{j\tau} + \sum_{i \in \mathcal{S}_j} \sum_{\tau=t+v_j}^{T} a_{ji} q_{i\tau} \qquad \begin{array}{l} j = 1,\dots,J \\ t = 1,\dots,T \end{array} \tag{3.36}$$

$$x_{jt} \leq \sum_{\tau=t}^{T} d_{j\tau} + \sum_{i \in \mathcal{S}_j} \sum_{\tau=t+v_j}^{T} a_{ji} q_{i\tau} \qquad \begin{array}{l} j = 1,\dots,J \\ t = 1,\dots,T \end{array} \tag{3.37}$$

$$y_{jt} \leq \sum_{\tau=t}^{T} d_{j\tau} + \sum_{i \in \mathcal{S}_j} \sum_{\tau=t+v_j}^{T} a_{ji} q_{i\tau} \qquad \begin{array}{l} j = 1,\dots,J \\ t = 1,\dots,T \end{array} \tag{3.38}$$

where the latter two inequalities assume that $q_{jt} > 0$ implies $q_{jt} \geq 1$ for $j = 1,\dots,J$ and $t = 1,\dots,T$.

Proof: Let j be the item under concern and period t be the focus of attention. If the right hand side is positive, the result is trivial due to (3.8). So, assume there is an optimum solution with $\sum_{\tau=t}^{T} d_{j\tau} + \sum_{i\in\mathcal{S}_j}\sum_{\tau=t+v_j}^{T} a_{ji}q_{i\tau} = 0$. Due to (3.2) in combination with (3.1) we get $q_{j\tau} = 0$ for all $\tau = t,\ldots,T$. With (3.6) there is an optimum solution with $y_{j\tau} = 0$ for all $\tau = t,\ldots,T$ which in turn implies $x_{j\tau} = 0$ for all $\tau = t,\ldots,T$ using (3.5). □

Valid constraint 9: Quite similar is a constraint which fixes setup state variables to zero if there is no future demand that is not met and if there is no initial inventory. Assume that if demand occurs, there is demand for at least one unit of an item.

$$
\begin{aligned}
y_{jt} \leq\ & y_{j(t+1)} & & j = 1,\ldots,J \\
& & & t = 1,\ldots,T-1 \\
& + \sum_{\tau=t}^{T} d_{j\tau} \\
& + \sum_{i\in\mathcal{S}_j}\sum_{\tau=t+v_j}^{T} a_{ji}q_{i\tau} \\
& - \sum_{\tau=t+2}^{T} q_{j\tau}
\end{aligned}
\tag{3.39}
$$

Proof: Let us concern about item j and period t. First, analyze the structure of the right hand side:

$$
\underbrace{y_{j(t+1)}}_{(I)} + \underbrace{\sum_{\tau=t}^{t+1} d_{j\tau} + \sum_{i\in\mathcal{S}_j}\sum_{\tau=t+v_j}^{t+v_j+1} a_{ji}q_{i\tau}}_{(II)}
$$
$$
+ \underbrace{\sum_{\tau=t+2}^{T} d_{j\tau} + \sum_{i\in\mathcal{S}_j}\sum_{\tau=t+v_j+2}^{T} a_{ji}q_{i\tau} - \sum_{\tau=t+2}^{T} q_{j\tau}}_{(III)}
$$

Part (I) represents the setup state in period $t+1$. Part (II) evaluates to a positive value if there is any demand in period t or period $t+1$. Eventually, part (III) represents future demand that is not met. If the right hand side is positive, the result is trivial due to

(3.8). So, assume an optimum solution where the right hand side evaluates to zero. Since $y_{j(t+1)} = 0$ no additional setup occurs in period t if $y_{jt} = 0$, too. For having no demand in periods t and $t+1$ to be met with production in t and $t+1$, respectively, and no future demand in periods $t+2$ to T which is not met an optimum solution would have $q_{jt} = q_{j(t+1)} = 0$ due to (3.2) in combination with (3.1). Hence, there is an optimum solution with $y_{jt} = 0$. □

3.3.8 Postprocessing

Given a particular instance, our goal is to heuristically find a (feasible) production plan with (sub–)optimum objective function value where finding a production plan means to determine the lot sizes and a schedule for the lots. Calculating the corresponding objective function value of a production plan is an easy task then.

In this subsection we will learn that once we have found a feasible solution represented by the matrix of production quantities q_{jt}, there are some degrees of freedom for the schedule. More precisely, a q_{jt}–matrix does not define a unique schedule — in our PLSP–MM–model formulation we thus introduced the setup state variables y_{jt} which determine a schedule.

A small (single–level) example should make things more clear. Suppose $J = 2$, $M = 1$, $T = 5$ and feasible production quantities as given in Table 3.10. Let the machine initially be set up for none of the items. Furthermore, assume $s_1 > s_2$. Since we do not need to know all the other parameters, they are not given here. Figure 3.9 shows three alternative feasible schedules.

q_{jt}	$t=1$	2	3	4	5
$j=1$		10		10	
$j=2$		10		10	

Table 3.10: A Matrix of Feasible Production Quantities

Note, all schedules differ only in the sequence of lots within the periods. Assuming the production quantities to be feasible, all

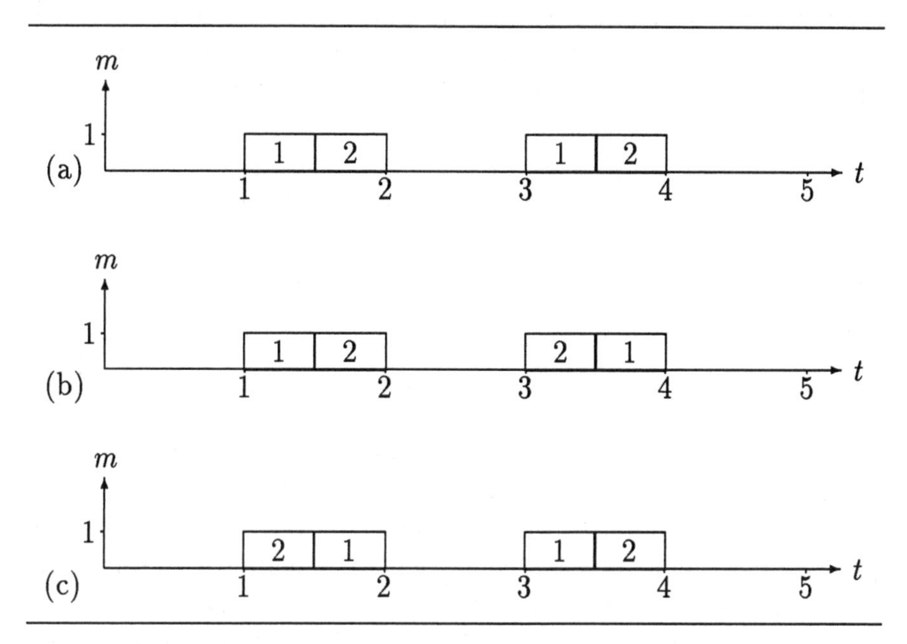

Figure 3.9: Three Feasible Schedules

three Gantt–charts thus define feasible production plans. Also note that holding costs are determined with the matrix of production quantities. Hence, all three schedules cause the same holding costs. But, if we consider setup costs, we can see that all three schedules cause different setup costs, namely, $2s_1 + 2s_2$ for schedule (a), $2s_1 + s_2$ for schedule (b), and $s_1 + 2s_2$ for schedule (c). By the way, no other schedule that corresponds to the quantities in Table 3.10 has a lower objective function value than schedule (c).

In general, we like to have a postprocessor which, once that we have found a feasible solution, determines a schedule with minimum total setup costs. o guarantee that the resulting schedule is feasible, the postprocessor should not modify the matrix of production quantities.[10] Under this assumption postprocessing the

[10]A formal statement of this problem emerges from the PLSP–MM–model formulation if production quantities q_{jt} are used as parameters and not as decision variables.

schedules for different machines does not interact. Basically, such a module could enumerate all sequences of lots within a period for all machines, thus enumerating $\sum_{m=1}^{M} 2^{\check{T}_m}$ schedules where

$$\check{T}_m \stackrel{def}{=} \sum_{t=1}^{T} \chi_2 \left(-1 + \sum_{j \in \mathcal{J}_m} \chi_2(q_{jt}) \right)$$

for all $m = 1, \ldots, M$ and χ_2 is an auxiliary function defined as

$$\chi_2(x) \stackrel{def}{=} \begin{cases} 1 & , \text{if } x > 0 \\ 0 & , \text{otherwise} \end{cases} . \tag{3.40}$$

Subsequently, we present a more efficient procedure.

We give here a functional specification of a postprocessor Ξ which, when applied to each machine, moves from period 1 to period t in a single pass and constructs the schedule with minimum total setup costs making use of a feasible q_{jt}–matrix. The output will be represented as a list of item indices. Each entry in the list corresponds to a positive entry in the matrix of production quantities. Hence, we have

$$\sum_{j \in \mathcal{J}_m} \sum_{t=1}^{T} \chi_2(q_{jt})$$

entries in the list representing the schedule for machine m. Reading the list from left to right uniquely defines a sequence of production quantities (and in combination with the q_{jt}–matrix a schedule, of course). For instance, applying the postprocessing phase to the example above should give $< 2 : 1 : 1 : 2 >_1$ representing the schedule for machine 1 in Figure 3.9–(c).[11]

Before we can give a formal definition of the postprocessor, let us introduce some notation for the sake of convenience. If machine m and period t are under concern, we have to evaluate $\sum_{j \in \mathcal{J}_m} \chi_2(q_{jt})$ using Σ_{mt} as a short–hand notation for it. Note,

[11]The formal definition of the postprocessor will need some operations on lists. We use $<>_m$ to denote an empty list, and a $\circ$–symbol to denote concatenation. For example, $< a : \ldots : z >_m \circ <>=< a : \ldots : z >_m$ and $< a : \ldots : w >_m \circ < x : y : z >_m=< a : \ldots : z >_m$.

since we consider the PLSP where at most one setup may take place per period, $\Sigma_{mt} \in \{0, 1, 2\}$. If $\Sigma_{mt} = 1$, let j^* be the unique item with $q_{j^*t} > 0$. If $\Sigma_{mt} = 2$, let $j_1^* < j_2^*$ be the items with non–zero production quantities. Furthermore, let j_{m0} be the item machine m is initially setup for, or $j_{m0} = 0$ if the machine m is set up for no item initially.

To apply the postprocessor Ξ to machine m, just evaluate

$$\Xi(m) \stackrel{def}{=} \Xi_1^m(j_{m0}, 1). \tag{3.41}$$

The function Ξ_1^m handles the case in which a period t is under concern and the setup state at the beginning of this period is fixed. This certainly happens to be in period 1 where the setup state at the beginning of that period is j_{m0}. Five cases may occur:
Case 1: $t > T$
Case 2: $\Sigma_{mt} = 0$ and $t \leq T$
Case 3: $\Sigma_{mt} = 1$ and $j^* = j$ and $t \leq T$
Case 4: $\Sigma_{mt} = 1$ and $j^* \neq j$ and $t \leq T$
Case 5: $\Sigma_{mt} = 2$ and $j_i^* = j$ $(i \in \{1, 2\})$ and $t \leq T$

Depending on the case that holds, Ξ_1^m is defined as follows:

$$\Xi_1^m(j,t) \stackrel{def}{=} \begin{cases} <> & \textit{Case 1} \\ \Xi_2^m(j, t+1) & \textit{Case 2} \\ < j > \circ\, \Xi_2^m(j^*, t+1) & \textit{Case 3} \\ < j^* > \circ\, \Xi_1^m(j^*, t+1) & \textit{Case 4} \\ < j : j_{(3-i)}^* > \circ\, \Xi_1^m(j_{(3-i)}^*, t+1) & \textit{Case 5} \end{cases} \tag{3.42}$$

First, there is the terminal case when all periods are considered. Second, if no production takes place in period t on machine m, no entry is added to the list which defines the schedule. A recursive call to Ξ_2^m passes the information that item j is the last item the machine is setup for. Third, there is the case where only one item j is produced on machine m in period t and machine m is already setup for this item. We then add the item index j to the list and proceed on with a call to Ξ_2^m. The fourth case is similar to the third, but this time the machine is not in the proper setup state. We thus add j^* to the list and call Ξ_1^m, because at the beginning of period $t + 1$ the machine is now set up for item j^*.

Finally, if there are two items to be produced, one of them must be item j. Otherwise, we would have two setups in period t which is not allowed. The sequence in which these items are produced must therefore have item j at its first position. At the end of period t and thus at the beginning of period $t+1$, the machine is set up for the second item then.

The function Ξ_2^m handles the case in which a period t is under concern and the setup state at the beginning of that period is not fixed yet. Again, we have five cases to distinguish.

Case 1: $t > T$
Case 2: $\Sigma_{mt} = 0$ and $t \leq T$
Case 3: $\Sigma_{mt} = 1$ and $t \leq T$
Case 4: $\Sigma_{mt} = 2$ and $j_i^* = j$ $(i \in \{1,2\})$ and $t \leq T$
Case 5: The above cases do not hold.

Ξ_2^m is defined as follows:

$$\Xi_2^m(j,t) \stackrel{def}{=} \begin{cases} <> & \textit{Case 1} \\ \Xi_2^m(j,t+1) & \textit{Case 2} \\ < j^* > \circ\, \Xi_2^m(j^*,t+1) & \textit{Case 3} \\ < j : j^*_{(3-i)} > \circ\, \Xi_1^m(j^*_{(3-i)},t+1) & \textit{Case 4} \\ \hat{\Xi}_1^m(j_1^*,j_2^*,t+1) & \textit{Case 5} \end{cases} \tag{3.43}$$

The first one is the terminal case. Case number two deals with situations in which nothing is to produce on machine m in period t. We simply move on with a recursive call to Ξ_2^m then. In case three, which is the case when only one item is to be manufactured, the item j^* is added to the list that defines the schedule. More interesting is case four where two items are to be produced on machine m in period t and one of them is the last item the machine is set up for. If this holds, it is clear that item j is to be scheduled first, because otherwise the setup costs for item j would be charged twice. Last, it may happen that two items are to be produced, but none of them is the one the machine is already set up for. A call to $\hat{\Xi}_1^m$ finds out the right sequence.

The function $\hat{\Xi}_1^m$ handles the case in which period t is under concern and the setup state at the beginning of that period is fixed. In contrast to the function Ξ_1^m, we employ $\hat{\Xi}_1^m$ if we are not quite sure which item the machine should be set up for at the beginning

of period t. Two alternatives do exist. In total we have seven cases to discriminate:
Case 1: $t > T$
Case 2: $\Sigma_{mt} = 0$ and $t \leq T$
Case 3: $\Sigma_{mt} = 1$ and $j^* = j_i$ $(i \in \{1, 2\})$ and $t \leq T$
Case 4: $\Sigma_{mt} = 1$ and $j_1 \neq j^* \neq j_2$ and $t \leq T$
Case 5: $\Sigma_{mt} = 2$ and $j_i^* = j_i \wedge j_{(3-i)}^* \neq j_{(3-i)}$ $(i \in \{1, 2\})$ and $t \leq T$
Case 6: $\Sigma_{mt} = 2$ and $j_1^* = j_1 \wedge j_2^* = j_2$ and $s_{j_1} \geq s_{j_2}$ and $t \leq T$
Case 7: $\Sigma_{mt} = 2$ and $j_1^* = j_1 \wedge j_2^* = j_2$ and $s_{j_1} < s_{j_2}$ and $t \leq T$

The function $\hat{\Xi}_1^m$ is defined as follows:

$$\hat{\Xi}_1^m(j_1, j_2, t) \stackrel{def}{=} \begin{cases} <> & \textit{Case 1} \\ \hat{\Xi}_2^m(j_1, j_2, t+1) & \textit{Case 2} \\ < j_{(3-i)} : j_i : j_i > \circ\, \Xi_2^m(j^*, t+1) & \textit{Case 3} \\ < j_1 : j_2 : j^* > \circ\, \Xi_1^m(j^*, t+1) & \textit{Case 4} \\ < j_{(3-i)} : j_i : j_i : j_{(3-i)}^* > & \\ \circ\, \Xi_1^m(j_{(3-i)}^*, t+1) & \textit{Case 5} \\ < j_2 : j_1 : j_1 : j_2 > \circ\, \Xi_1^m(j_2^*, t+1) & \textit{Case 6} \\ < j_1 : j_2 : j_2 : j_1 > \circ\, \Xi_1^m(j_1^*, t+1) & \textit{Case 7} \end{cases} \tag{3.44}$$

Many cases may occur. Of course, there is the terminal case. If nothing is to be produced in period t, we can still not be sure in which sequence the items that are kept in memory are to be scheduled. However, we know that the setup state of machine m at the end of period t is not fixed yet, and thus we call $\hat{\Xi}_2^m$. The third case deals with situations in which one item is to be produced in period t and in which this single item equals one of those that are remembered. Hence, we build a lot for that item which is the cheapest possible sequence. If the item that is to be produced is not equal to one of those that are pending to be added to the list, case four holds. The sequence in which the two pending items are scheduled can be chosen arbitrarily then. Case five is the case where two items are to be scheduled in period t, but only one of them equals one of the pending items. The sequence in which the items are scheduled is the one which causes the least setup costs. Cases six and seven describe what to do if the two pending items are to be produced on machine m in period t again. A lot is then built for the item with the highest setup cost.

The function $\hat{\Xi}_2^m$ handles the case in which period t is under concern and the setup state at the beginning of that period is not fixed. Also, we face a pending sequence decision from earlier periods which is in difference to Ξ_2^m. The function $\hat{\Xi}_2^m$ is defined as follows:

$$\hat{\Xi}_2^m(j_1, j_2, t) \stackrel{def}{=} \begin{cases} \ldots & \text{, see } \textit{Cases 1 to 7} \text{ of } \hat{\Xi}_1^m \\ < j_1 : j_2 > & \\ \circ\ \hat{\Xi}_1^m(j_1^*, j_2^*, t+1) & \text{, otherwise} \end{cases} \tag{3.45}$$

This definition is almost equal to the definition of $\hat{\Xi}_1^m$, i.e. we have the same seven cases as before. The only difference is an eighth case which applies when two items are to be produced in period t and none of them equals any of the pending items.

To illustrate the formulae, let us run a postprocessing phase for the example above where production quantities are given in Table 3.10. Remember, $j_{10} = 0$ and $s_1 > s_2$.

$$\begin{aligned} \Xi(1) &= \Xi_1^1(0,1) \\ &= \Xi_2^1(0,2) \\ &= \hat{\Xi}_1^1(1,2,3) \\ &= \hat{\Xi}_2^1(1,2,4) \\ &= < 2:1:1:2 >_1 \circ\ \Xi_1^1(2,5) \\ &= < 2:1:1:2 >_1 \circ\ \Xi_2^1(2,6) \\ &= < 2:1:1:2 >_1 \circ <>_1 \\ &= < 2:1:1:2 >_1 \end{aligned}$$

This result indeed corresponds to the schedule given in Figure 3.9–(c).

3.4 Parallel Machines

Lot sizing and scheduling with parallel machines (PLSP–PM) introduces a new degree of freedom into the planning problem. In contrast to lot sizing and scheduling with multiple machines we now have no fixed machine assignments for the items. In other words, it is a priori not clear on which machine items are produced. We assume that some (maybe all) machines are capable to produce a particular item. The multi–machine case obviously is a special case of the parallel–machine case, because the number of machines to choose among is one for each item. This actually means that there is no choice. Furthermore, we assume the most general case that is the case of heterogeneous machines. This is to say, that the production of the same item requires different amounts of capacity if different machines are used.

Table 3.11 defines the decision variables which are new or redefined. Likewise, Table 3.12 gives the parameters. A MIP–model formulation can now be presented to give a precise problem statement.

Symbol	Definition
q_{jmt}	Production quantity for item j on machine m in period t.
x_{jmt}	Binary variable which indicates whether a setup for item j occurs on machine m in period t ($x_{jmt} = 1$) or not ($x_{jmt} = 0$).
y_{jmt}	Binary variable which indicates whether machine m is set up for item j at the end of period t ($y_{jmt} = 1$) or not ($y_{jmt} = 0$).

Table 3.11: New Decision Variables for the PLSP–PM

$$\min \sum_{m=1}^{M} \sum_{j \in \mathcal{J}_m} \sum_{t=1}^{T} s_{jm} x_{jmt} \tag{3.46}$$

$$+ \sum_{j=1}^{J} \sum_{t=1}^{T} h_j I_{jt}$$

subject to

Symbol	Definition
$\mathcal{J}_m$	Set of all items that share the machine m, i.e. $\mathcal{J}_m \stackrel{def}{=} \{j \in \{1,\ldots,J\} \mid m \in \mathcal{M}_j\}$.
$\mathcal{M}_j$	Set of all machines that are capable to produce item j, i.e. $\mathcal{M}_j \stackrel{def}{=} \{m \in \{1,\ldots,M\} \mid p_{jm} < \infty\}$.
p_{jm}	Capacity needs for producing one unit of item j on machine m. Its value is ∞ if machine m cannot be used to produce item j.
s_{jm}	Non–negative setup cost for item j on machine m.
y_{jm0}	Initial setup state.

Table 3.12: New Parameters for the PLSP–PM

$$I_{jt} = I_{j(t-1)} + \sum_{m \in \mathcal{M}_j} q_{jmt} - d_{jt} - \sum_{i \in \mathcal{S}_j} \sum_{m \in \mathcal{M}_i} a_{ji} q_{imt} \qquad j = 1,\ldots,J;\ t = 1,\ldots,T \tag{3.47}$$

$$I_{jt} \geq \sum_{i \in \mathcal{S}_j} \sum_{m \in \mathcal{M}_i} \sum_{\tau=t+1}^{\min\{t+v_j,T\}} a_{ji} q_{im\tau} \qquad j = 1,\ldots,J;\ t = 0,\ldots,T-1 \tag{3.48}$$

$$\sum_{j \in \mathcal{J}_m} y_{jmt} \leq 1 \qquad m = 1,\ldots,M;\ t = 1,\ldots,T \tag{3.49}$$

$$x_{jmt} \geq y_{jmt} - y_{jm(t-1)} \qquad j = 1,\ldots,J;\ m \in \mathcal{M}_j;\ t = 1,\ldots,T \tag{3.50}$$

$$p_{jm} q_{jmt} \leq C_{mt}(y_{jm(t-1)} + y_{jmt}) \qquad j = 1,\ldots,J;\ m \in \mathcal{M}_j;\ t = 1,\ldots,T \tag{3.51}$$

$$\sum_{j \in \mathcal{J}_m} p_{jm} q_{jmt} \leq C_{mt} \qquad m = 1,\ldots,M;\ t = 1,\ldots,T \tag{3.52}$$

$$y_{jmt} \in \{0,1\} \qquad j = 1,\ldots,J;\ m \in \mathcal{M}_j;\ t = 1,\ldots,T \tag{3.53}$$

$$I_{jt} \geq 0 \qquad \begin{array}{l} j = 1, \ldots, J \\ t = 1, \ldots, T \end{array} \tag{3.54}$$

$$q_{jmt}, x_{jmt} \geq 0 \qquad \begin{array}{l} j = 1, \ldots, J \\ m \in \mathcal{M}_j \\ t = 1, \ldots, T \end{array} \tag{3.55}$$

The meaning of (3.46) to (3.55) closely relates to (3.1) to (3.9) and thus needs no further explanation. However, it is remarkable to note that lots of the same item may be produced on different machines in the same period. Splitting lots for concurrent production on different machines may be necessary to find feasible (and optimum) solutions.

3.5 Multiple Resources

For lot sizing and scheduling with multiple resources (PLSP–MR) we assume that each item requires several resources at a time. Manufacturing an item needs all corresponding resources to be in the right setup state. Thus, the model has to guarantee that, if an item is produced in a period, all required resources are in the proper setup state either at the beginning or at the end of the period. The multi–machine problem is a special case of multiple resources. In the former case items require only one resource to be produced.

Table 3.13 introduces the new decision variables and Table 3.14 gives the new parameters that are needed to extend the multi–machine problem. A MIP–model formulation using this notation can now be presented.

$$\min \sum_{m=1}^{M} \sum_{j \in \mathcal{J}_m} \sum_{t=1}^{T} s_{jm} x_{jmt} \tag{3.56}$$

$$+ \sum_{j=1}^{J} \sum_{t=1}^{T} h_j I_{jt}$$

subject to

Symbol	Definition
q^B_{jt}	Production quantity for item j in period t where the required resources are properly set up at the beginning of period t.
q^E_{jt}	Production quantity for item j in period t where the required resources are properly set up at the end of period t.
x_{jmt}	Binary variable which indicates whether a setup for item j occurs on resource m in period t ($x_{jmt} = 1$) or not ($x_{jmt} = 0$).
y_{jmt}	Binary variable which indicates whether resource m is set up for item j at the end of period t ($y_{jmt} = 1$) or not ($y_{jmt} = 0$).

Table 3.13: New Decision Variables for the PLSP–MR

$$I_{jt} = I_{j(t-1)} + q^B_{jt} + q^E_{jt} - d_{jt} - \sum_{i \in \mathcal{S}_j} a_{ji}(q^B_{it} + q^E_{it}) \qquad \begin{array}{l} j = 1, \ldots, J \\ t = 1, \ldots, T \end{array} \tag{3.57}$$

$$I_{jt} \geq \sum_{i \in \mathcal{S}_j} \sum_{\tau = t+1}^{\min\{t+v_j, T\}} a_{ji}(q^B_{i\tau} + q^E_{i\tau}) \qquad \begin{array}{l} j = 1, \ldots, J \\ t = 0, \ldots, T-1 \end{array} \tag{3.58}$$

$$\sum_{j \in \mathcal{J}_m} y_{jmt} \leq 1 \qquad \begin{array}{l} m = 1, \ldots, M \\ t = 1, \ldots, T \end{array} \tag{3.59}$$

$$x_{jmt} \geq y_{jmt} - y_{jm(t-1)} \qquad \begin{array}{l} j = 1, \ldots, J \\ m \in \mathcal{M}_j \\ t = 1, \ldots, T \end{array} \tag{3.60}$$

$$p_{jm} q^B_{jt} \leq C_{mt} y_{jm(t-1)} \qquad \begin{array}{l} j = 1, \ldots, J \\ m \in \mathcal{M}_j \\ t = 1, \ldots, T \end{array} \tag{3.61}$$

$$p_{jm} q^E_{jt} \leq C_{mt} y_{jmt} \qquad \begin{array}{l} j = 1, \ldots, J \\ m \in \mathcal{M}_j \\ t = 1, \ldots, T \end{array} \tag{3.62}$$

Symbol	Definition
$\mathcal{J}_m$	Set of all items that share the resource m, i.e. $\mathcal{J}_m \stackrel{def}{=} \{j \in \{1,\ldots,J\} \mid p_{jm} < \infty\}$.
$\mathcal{M}_j$	Set of all resources that are needed to produce item j, i.e. $\mathcal{M}_j \stackrel{def}{=} \{m \in \{1,\ldots,M\} \mid p_{jm} < \infty\}$.
p_{jm}	Capacity needs for resource m for producing one unit of item j. Its value is ∞ if item j does not require resource m.
s_{jm}	Non–negative setup cost for item j on resource m.
y_{jm0}	Initial setup state.

Table 3.14: New Parameters for the PLSP–MR

$$\sum_{j \in \mathcal{J}_m} p_{jm}(q_{jt}^B + q_{jt}^E) \leq C_{mt} \qquad m = 1,\ldots,M; \; t = 1,\ldots,T \tag{3.63}$$

$$y_{jmt} \in \{0,1\} \qquad j = 1,\ldots,J; \; m \in \mathcal{M}_j; \; t = 1,\ldots,T \tag{3.64}$$

$$I_{jt}, q_{jt}^B, q_{jt}^E \geq 0 \qquad j = 1,\ldots,J; \; t = 1,\ldots,T \tag{3.65}$$

$$x_{jmt} \geq 0 \qquad j = 1,\ldots,J; \; m \in \mathcal{M}_j; \; t = 1,\ldots,T \tag{3.66}$$

This MIP–formulation is a straightforward extension of the PLSP–MM–model and thus needs no elaborate discussion. However, an important aspect should be mentioned briefly. Due to the decision variables q_{jt}^B and q_{jt}^E we explicitly take into account the sequence of two different lots in a period. This is redundant in the PLSP–MM–model since the setup state variables y_{jmt} uniquely define sequences. But now, we must make sure that all resources needed to produce an item j are in the right setup state at a time. Constraints (3.61) and (3.62) guarantee so. Noteworthy to say that postprocessing as described above cannot be done here since the schedules of different resources do interact.

3.6 Partially Renewable Resources

For lot sizing and scheduling with partially renewable resources (PLSP–PRR) we assume that each item requires one resource for which a setup state has to be taken into account. This is equivalent to the multi–machine case. In addition, each item may require several scarce resources for which no setup state has to be taken into account. Moreover, since we have small periods representing shifts or hours for instance, we assume that these additional resources have capacity limits given per interval of periods.[12] Suppose for example, that periods represent shifts (e.g. 10 shifts per week) and resources are renewed once a week. Since they are not renewed in every period they are called partially renewable. Capacity limits are then given per week and not per shift. The case where capacity limits are given per period is of course a special case. Such resources would be called renewable. The one for which a setup state is taken into account is a renewable resource.

To understand the subsequent MIP–model formulation, see Table 3.15 for new parameters. The decision variables equal those of the PLSP–MM–model.

$$\min \sum_{j=1}^{J} \sum_{t=1}^{T} (s_j x_{jt} + h_j I_{jt}) \tag{3.67}$$

subject to

$$I_{jt} = I_{j(t-1)} + q_{jt} - d_{jt} - \sum_{i \in \mathcal{S}_j} a_{ji} q_{it} \qquad \begin{array}{l} j = 1, \ldots, J \\ t = 1, \ldots, T \end{array} \tag{3.68}$$

$$I_{jt} \geq \sum_{i \in \mathcal{S}_j} \sum_{\tau = t+1}^{\min\{t+v_j, T\}} a_{ji} q_{i\tau} \qquad \begin{array}{l} j = 1, \ldots, J \\ t = 0, \ldots, T-1 \end{array} \tag{3.69}$$

$$\sum_{j \in \mathcal{J}_m} y_{jt} \leq 1 \qquad \begin{array}{l} m = 1, \ldots, M \\ t = 1, \ldots, T \end{array} \tag{3.70}$$

[12]A more general point of view would be to consider arbitrary sets of periods instead of intervals of periods. However, this is not done here.

Symbol	Definition
$\tilde{C}_{mt}$	Available capacity of the partially renewable resource m in interval t.
$\tilde{\mathcal{J}}_m$	Set of all items that share the partially renewable resource m, i.e. $\tilde{\mathcal{J}}_m \stackrel{def}{=} \{j \in \{1, \ldots, J\} \mid \tilde{p}_{jm} < \infty\}$.
$\tilde{L}$	Length of an interval in number of periods. All intervals are assumed to have equal length.
$\tilde{M}$	Number of partially renewable resources.
$\tilde{\mathcal{M}}_j$	Set of all partially renewable resources that are needed to produce item j, i.e. $\tilde{\mathcal{M}}_j \stackrel{def}{=} \left\{m \in \{1, \ldots, \tilde{M}\} \mid \tilde{p}_{jm} < \infty\right\}$.
$\tilde{p}_{jm}$	Capacity needs for the partially renewable resource m for producing one unit of item j. Its value is ∞ if item j does not require resource m.
$\tilde{T}$	Number of intervals. We assume $\tilde{L} \cdot \tilde{T} = T$ and $[(i-1)\tilde{L}+1, i\tilde{L}]$ be the i–th interval of periods where $i = 1, \ldots, \tilde{T}$. Note, a period $t \in \{1, \ldots, T\}$ belongs to the interval $i = \left\lceil \frac{t}{\tilde{L}} \right\rceil$.

Table 3.15: New Parameters for the PLSP–PRR

$$x_{jt} \geq y_{jt} - y_{j(t-1)} \qquad \begin{array}{l} j = 1, \ldots, J \\ t = 1, \ldots, T \end{array} \tag{3.71}$$

$$p_j q_{jt} \leq C_{m_j t}(y_{j(t-1)} + y_{jt}) \qquad \begin{array}{l} j = 1, \ldots, J \\ t = 1, \ldots, T \end{array} \tag{3.72}$$

$$\sum_{j \in \mathcal{J}_m} p_j q_{jt} \leq C_{mt} \qquad \begin{array}{l} m = 1, \ldots, M \\ t = 1, \ldots, T \end{array} \tag{3.73}$$

$$\sum_{j \in \tilde{\mathcal{J}}_m} \sum_{\tau=(t-1)\tilde{L}+1}^{t\tilde{L}} \tilde{p}_{jm} q_{j\tau} \leq \tilde{C}_{mt} \qquad \begin{array}{l} m = 1, \ldots, \tilde{M} \\ t = 1, \ldots, \tilde{T} \end{array} \tag{3.74}$$

$$y_{jt} \in \{0, 1\} \qquad \begin{array}{l} j = 1, \ldots, J \\ t = 1, \ldots, T \end{array} \tag{3.75}$$

$$I_{jt}, q_{jt}, x_{jt} \geq 0 \qquad \begin{array}{c} j = 1, \ldots, J \\ t = 1, \ldots, T \end{array} \qquad (3.76)$$

All constraints but (3.74) equal those of the PLSP–MM–model and thus need no explanation again. The new set of restrictions (3.74) represents the capacity limits of the partially renewable resources. They make sure that the sum of capacity demands for resource m in the t–th interval of periods does not exceed the capacity limit.

Chapter 4

Instance Generation

As we have learned, the problem that we are concerned about has not been treated elsewhere. Hence, there is neither an established multi–level test–bed nor an instance generator available. Section 4.1 therefore introduces a parameter controlled instance generator (APCIG) which allows to create multi–level lot sizing (and scheduling) instances systematically. Due to preliminary computational experience, we know that certain parameters of the PLSP have an impact on the performance of solution methods. This is not a very surprising result, of course. But, to gain a better understanding in what makes instances hard to solve and what method should be chosen for what instance, a (full) factorial design is used rather than a randomized design. For a guideline for designing test–beds, performing computational studies, and reporting the results[1] we refer to [BaGoKeReSt95, Hoo95]. Section 4.2 presents an experimental design for the PLSP–MM. Section 4.3 relates the speed of different computers to allow fair comparisons with other platforms than those used for our tests.

4.1 Methods

Let us consider the PLSP–MM first. Subsequent subsections will explain what changes when generating instances for extensions.

[1]Statistical methods for data analysis are described in [Ott93].

Throughout the text we make use of some short–hand notation. To draw a real–valued random variable with uniform distribution out of an interval $[a, b]$ we write $\in RRAND[a, b]$. Analogously, $\in IRAND[a, b]$ is used to draw an integer–valued random variable. Generating uniformly distributed random numbers is based on the random number generator described in [Sch79]. If x is real–valued, $\lfloor x \rfloor$ ($\lceil x \rceil$) denotes the greatest (smallest) integer value that is less (greater) than or equal to x.

All PLSP–MM parameters out of J, M, and T (which are user input respecting $M \leq J$) are generated at random. Straightforwardly, we choose

$I_{j0} \in IRAND[INITINV_{min}, INITINV_{max}]$
where $INITINV_{max} \geq INITINV_{min} \geq 0$ are user input,

$m_j \in IRAND[1, M]$
where $\mathcal{J}_m \neq \emptyset$ for all $m = 1, \ldots, M$ must hold,

$p_j \in RRAND[CAPNEED_{min}, CAPNEED_{max}]$
where $CAPNEED_{max} \geq CAPNEED_{min} > 0$ are user input,

$v_j \in IRAND[LEADTIME_{min}, LEADTIME_{max}]$
where $LEADTIME_{max} \geq LEADTIME_{min} > 0$ are user input,

$y_{j0} \in IRAND[0, 1]$
where $\sum_{j \in \mathcal{J}_m} y_{j0} \leq 1$ for all $m = 1, \ldots, M$ must hold.

The remaining parameters for the PLSP–MM need a more sophisticated strategy to be generated randomly.

4.1.1 Generation of Gozinto–Structures

The input parameters for generating acyclic gozinto–structures are given in Table 4.1.

In the sequel we confine ourself to general gozinto–structures ($TYPE_G = general$) in order to avoid too much technical detail. The generation of linear, assembly, or divergent gozinto–structures is roughly the same. Some hints for what changes is spread in the text and should be sufficient.

Symbol	Definition
$[ARC_{min}, ARC_{max}]$	Interval of gozinto–factors where $ARC_{max} \geq ARC_{min} \geq 0$. To generate single–level instances one may choose $ARC_{min} = ARC_{max} = 0$.
$COMPLEXITY$	Complexity of the gozinto–structure.
$DEPTH$	Largest low level code of items in the gozinto–structures.
$ITEMS(0)$	Number of end items.
J	Number of items, where $ITEMS(0) + DEPTH \leq J$.
$MAXITER_G$	Maximum number of iterations where $MAXITER_G \geq N_G$.
N_G	Number of different gozinto–structures to be created.
$PRIVAL(llc)$	A priority value for $llc = 1, \ldots, DEPTH$ which is used to define a discrete probability function that helps to shape the gozinto–structure where $PRIVAL(llc) > 0$ for all $llc = 1, \ldots, DEPTH$.
$TYPE_G$	Type of the gozinto–structure, i.e. $TYPE_G \in \{linear, assembly, divergent, general\}$.

Table 4.1: Parameters for Generating Gozinto–Structures

Once the generation of gozinto–structures is started, we first decide how many items should have what low level code. Let $ITEMS(llc)$ where $llc = 1, \ldots, DEPTH$ denote the number of items with low level code llc. We must guarantee $ITEMS(llc) \geq 1$ for all $llc = 1, \ldots, DEPTH$ to come up with a structure of the desired depth. Such a structure must exist since $ITEMS(0) + DEPTH \leq J$ holds. At the end

$$\sum_{llc=0}^{DEPTH} ITEMS(llc) = J \tag{4.1}$$

must hold. Low level codes are assigned at random using a discrete probability function which is defined on the basis of $PRIVAL(llc)$ for all $llc = 1, \ldots, DEPTH$. Formally, if $Z \in RRAND[0, 1]$ then

$$level(Z) \stackrel{def}{=} \min \left\{ llc \in \{1, \ldots, DEPTH\} \mid \frac{\sum_{n=1}^{llc} PRIVAL(n)}{\sum_{n=1}^{DEPTH} PRIVAL(n)} \geq Z \right\} \tag{4.2}$$

is the low level code to choose next. The piece of code in Table 4.2 gives a precise definition of what is done.

$ITEMS(llc) := 1$ for all $llc = 1, \ldots, DEPTH$.
while $\sum_{llc=0}^{DEPTH} ITEMS(llc) < J$
    choose $Z \in RRAND[0, 1]$.
    $ITEMS(level(Z)) := ITEMS(level(Z)) + 1$.

Table 4.2: Method to Construct a Gozinto–Structure, Part 1

Figure 4.1 shows a possible outcome for $J = 6$, $ITEMS(0) = 1$, and $DEPTH = 2$.

If linear or divergent gozinto–structures shall be constructed, we must respect $ITEMS(llc) \leq ITEMS(llc - 1)$ for all $llc = 1, \ldots, DEPTH$ and keep this condition true during execution.

In the next step, we construct a gozinto–structure where exactly $ITEMS(llc)$ items have a low level code $llc = 0, \ldots, DEPTH$.

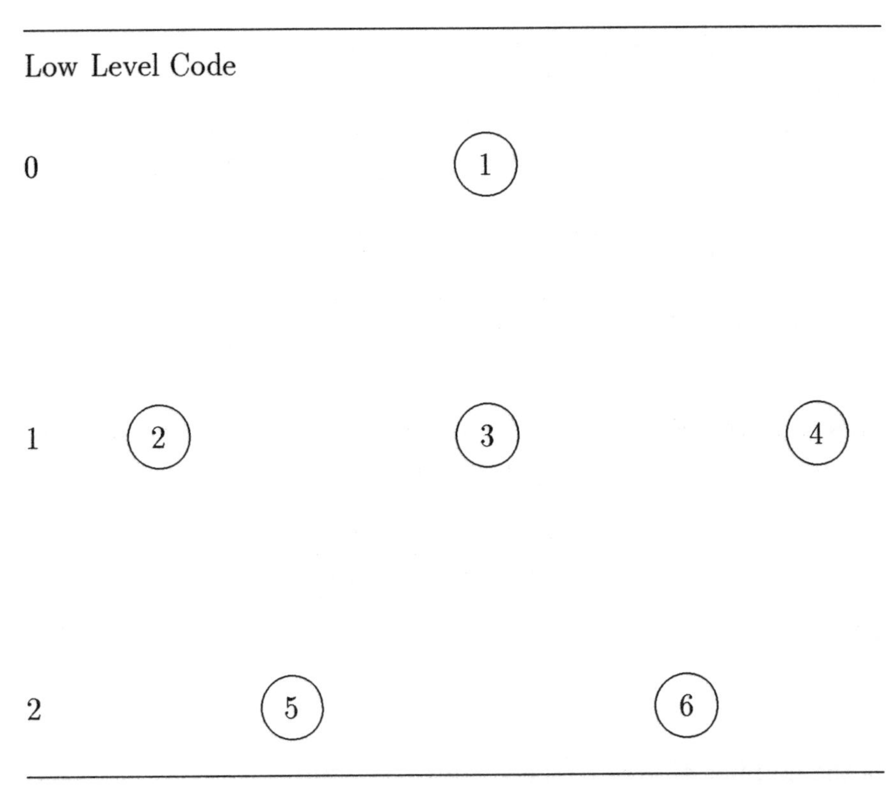

Figure 4.1: A Possible Assignment of Low Level Codes to Items

We will use the minimum number of arcs to do so. Without loss of generality we assume that items that should have a low level code llc are numbered consecutively from $1+\sum_{n=0}^{llc-1} ITEMS(n)$ to $\sum_{n=0}^{llc} ITEMS(n)$ (see Figure 4.1). This, by the way, guarantees a technological ordering. The basic idea now is simple. Just introduce exactly one arc pointing from an item that should have a low level code llc where $llc = 1, \ldots, DEPTH$ to any item that should have a low level code $llc - 1$. Figure 4.2 shows a possible result for the example where $ARC_{min} = 1$ and $ARC_{max} = 2$. More formally, this part of the construction of a gozinto–structure can be described as in Table 4.3.

The resulting gozinto–structure meets its specification with a minimum number of arcs, i.e. $J - ITEMS(0)$ arcs. Additional

$a_{ji} = 0$ for all $j, i = 1, \ldots, J$.
for $llc = 1$ to $llc = DEPTH$
    for $j = 1 + \sum_{n=0}^{llc-1} ITEMS(n)$ to $j = \sum_{n=0}^{llc} ITEMS(n)$
        choose $i \in IRAND[\ 1 + \sum_{n=0}^{llc-2} ITEMS(n),$
            $\sum_{n=0}^{llc-1} ITEMS(n)]$.
        $a_{ji} := IRAND[ARC_{min}, ARC_{max}]$.

Table 4.3: Method to Construct a Gozinto–Structure, Part 2

arcs can now be introduced in order to make the structure more "complex". A measure of complexity can be defined as[2]

$$\mathcal{C} \stackrel{def}{=} \frac{numarcs - minarcs}{maxarcs - minarcs} \tag{4.3}$$

where

$$numarcs \stackrel{def}{=} \sum_{j=1}^{J} \sum_{i=1}^{j-1} \chi_2(a_{ji}) \tag{4.4}$$

is the number of arcs in the gozinto–structure,

$$minarcs \stackrel{def}{=} J - ITEMS(0) \tag{4.5}$$

is the minimum number of arcs in a gozinto–structure with J items given the number $ITEMS(0)$ of end items, and

$$maxarcs \stackrel{def}{=} \sum_{llc=0}^{DEPTH-1} \left(ITEMS(llc) \sum_{n=llc+1}^{DEPTH} ITEMS(n) \right) \tag{4.6}$$

is the maximum number of arcs in a gozinto–structure with J items given the number $ITEMS(llc)$ of items per low level for $llc = 0, \ldots, DEPTH$. The complexity $\mathcal{C}$ of a multi–level gozinto–structure is a real value where $0 \leq \mathcal{C} \leq 1$ holds. Roughly speaking, it is a measure that expresses how many arcs could be removed while keeping the same number of items per low level, and it relates

[2]See (3.40) for a definition of the auxiliary function χ_2.

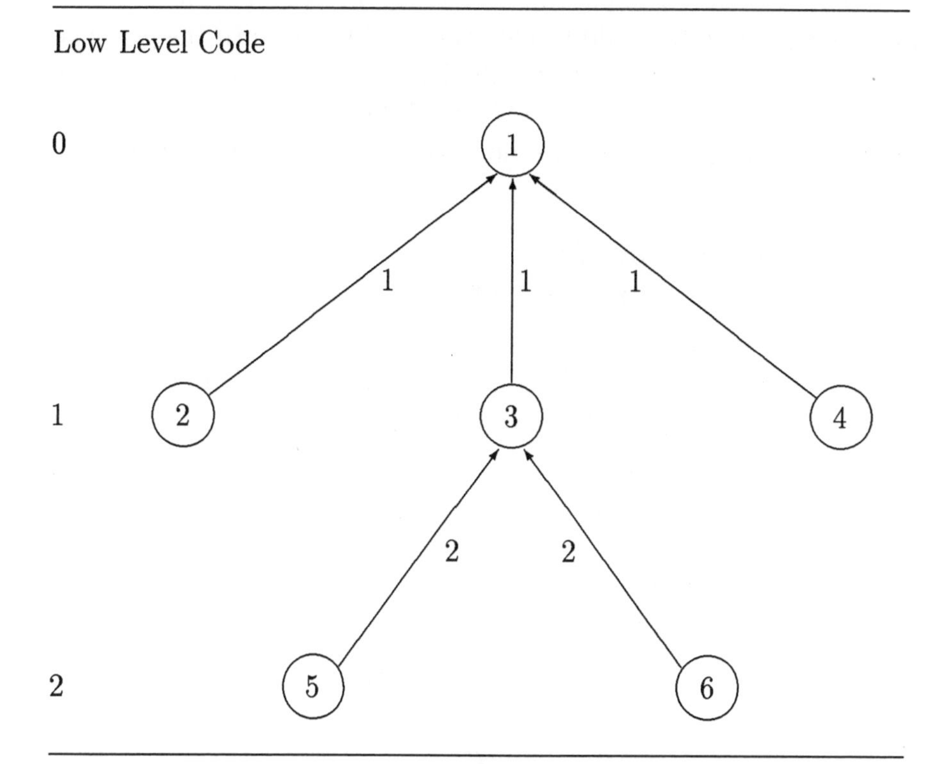

Figure 4.2: A Gozinto–Structure with a Minimum Number of Arcs

this value to the maximum additional number of arcs that could exist. $\mathcal{C}$ evaluates to zero if the gozinto–structure under concern has the minimum number of arcs (see Figure 4.2 for example), and its value is one if no further arcs can be introduced without changing the number of items per low level.[3]

[3]Discussions on complexity measures for gozinto–structures are rather undone. Only one author, namely Collier [Col81], has introduced a so–called degree of commonality index

$$\mathcal{C}' \stackrel{def}{=} \frac{\sum_{j=1+ITEMS(0)}^{J} \sum_{i=1}^{j-1} \sum_{k=1}^{ITEMS(0)} \chi_2\left(\chi_2(id_{ki})a_{ji}\right)}{J - ITEMS(0)}.$$

We do not use this definition, because we feel that the fact that the value of $\mathcal{C}'$ has no upper bound, which is valid for all gozinto–structures whatsoever,

This is where the parameter $COMPLEXITY$ comes in. We now keep on adding additional arcs one by one until the gozinto–structure has the desired complexity. Figure 4.3 gives a possible result for the example if $COMPLEXITY = 0.5$ is chosen ($\mathcal{C} = 0.5$, too). More precisely, we do what is described in Table 4.4.

---

while $\mathcal{C} < COMPLEXITY$
  choose $j \in IRAND[1 + ITEMS(0), J]$
    where $\sum_{i=1}^{j-1} \chi_2(a_{ji}) < \sum_{n=0}^{llc_j-1} ITEMS(n)$.
  choose $i \in IRAND[1, \sum_{n=0}^{llc_j-1} ITEMS(n)]$ where $a_{ji} = 0$.
  $a_{ji} := IRAND[ARC_{min}, ARC_{max}]$.

Table 4.4: Method to Construct a Gozinto–Structure, Part 3

After this procedure terminates, we have a gozinto–structure with complexity $\mathcal{C}$ where

$$\begin{array}{c} COMPLEXITY \\ \leq \\ \mathcal{C} \\ < \\ COMPLEXITY + \frac{1}{maxarcs - minarcs} \end{array}$$

holds.

Note, linear and assembly gozinto–structures always have a complexity $\mathcal{C} = 0$. In general, divergent structures have a maximum complexity $\mathcal{C} < 1$.

Since we want to have N_G different gozinto–structures, we simply iterate the whole procedure over and over again each time starting from scratch, i.e. starting with determining $ITEMS(llc)$ for all $llc = 1, \ldots, DEPTH$. This loop is halted, if N_G different gozinto–structures are found or if $MAXITER_G$ iterations

---

is less graphic than our definition. A value of $\mathcal{C}$ close to one indicates "many arcs" which means many predecessor relationships between items. But what does a $\mathcal{C}'$–value of, say, 1.5 mean? Since we use the complexity as a user defined parameter for generating gozinto–structures, the user should have a good feeling for the impact of changing its value.

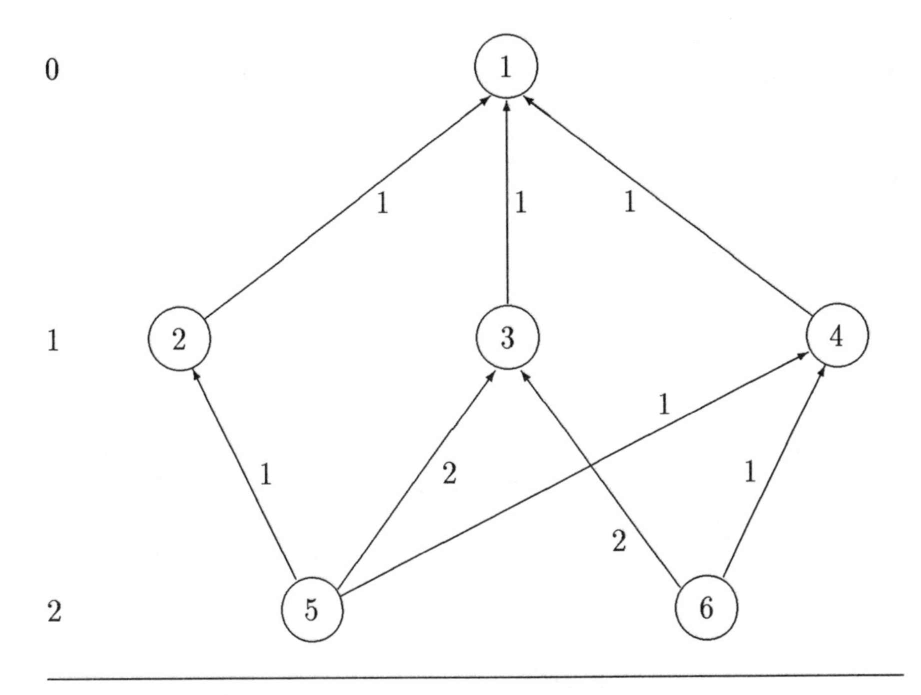

Figure 4.3: A Gozinto–Structure with $\mathcal{C} = 0.5$

are performed. The latter condition guarantees that the generation of gozinto–structures terminates even if N_G is greater than the number of different gozinto–structures which do exist while meeting the specification. Testing whether a generated gozinto–structure is new or has already been constructed during earlier iterations seems to be an easy task at first sight. One might guess that comparing the a_{ji}–matrices is sufficient.[4] Consider Figure 4.4 to see another gozinto–structure that could have been computed following the above lines. Both, Figure 4.3 and Figure 4.4 show isomorphic gozinto–structures. Since simple relabeling of the items

[4]If two adjacency matrices are of a different size such a comparison is, of course, sufficient.

makes both identical, we wish to identify them as the same structure.

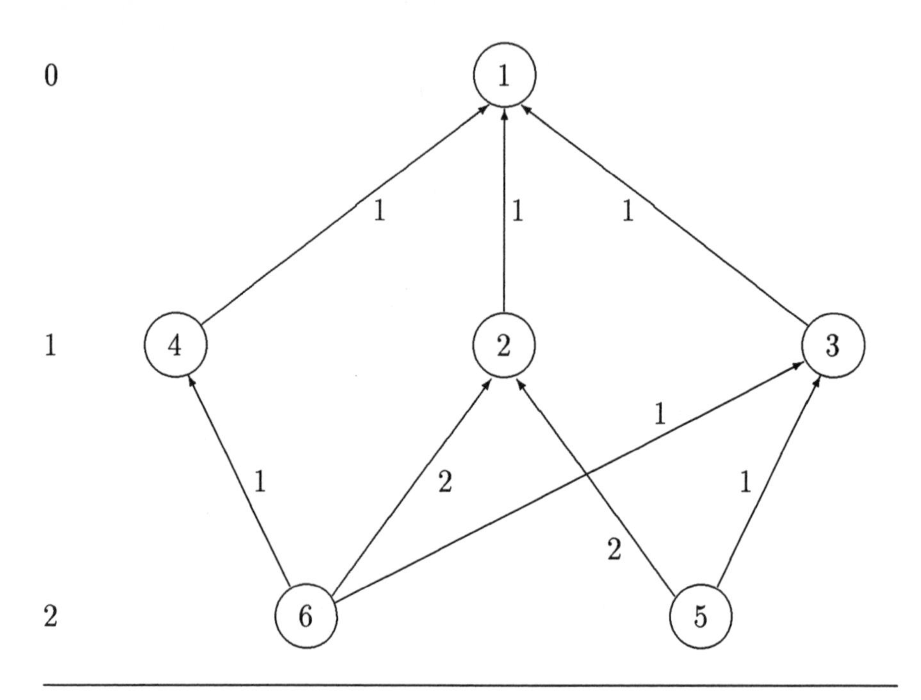

Figure 4.4: An Isomorphic Gozinto–Structure

Having a look at the a_{ji}–matrices reveals that both can hardly be identified as representing the same structure. We give the matrix for Figure 4.3 at the left hand side and the matrix that corresponds to Figure 4.4 at the right hand side:

$$\begin{pmatrix} 0 & 0 & 0 & 0 & 0 & 0 \\ 1 & 0 & 0 & 0 & 0 & 0 \\ 1 & 0 & 0 & 0 & 0 & 0 \\ 1 & 0 & 0 & 0 & 0 & 0 \\ 0 & 1 & 2 & 1 & 0 & 0 \\ 0 & 0 & 2 & 1 & 0 & 0 \end{pmatrix} \overset{?}{\Longleftrightarrow} \begin{pmatrix} 0 & 0 & 0 & 0 & 0 & 0 \\ 1 & 0 & 0 & 0 & 0 & 0 \\ 1 & 0 & 0 & 0 & 0 & 0 \\ 1 & 0 & 0 & 0 & 0 & 0 \\ 0 & 2 & 1 & 0 & 0 & 0 \\ 0 & 2 & 1 & 1 & 0 & 0 \end{pmatrix}$$

A more sophisticated method thus needs to be employed to test whether or not two adjacency matrices represent the same gozinto–structure. If the values $ITEMS(llc)$ of the two gozinto–structures differ for at least one $llc = 0, \ldots, DEPTH$, it is obvious that the two matrices cannot represent the same gozinto–structure. Otherwise, the trick to check if two adjacency matrices represent the same gozinto–structure is to consider the corresponding incidence matrices, find pairs of so–called equivalent rows, and delete these. If this yields all rows deleted, the two incidence matrices and hence both adjacency matrices represent the same gozinto–structure. If not, they represent different structures. Before we begin to explain what equivalent rows are, let us recall the notion of an incidence matrix $(\hat{a}_{ji})_{j,i=1,\ldots,J}$ from graph theory:

$$\hat{a}_{ji} = \begin{cases} a_{ji}, & \text{, if } i \in \mathcal{S}_j \\ -a_{ij}, & \text{, if } i \in \mathcal{P}_j \end{cases} \qquad j, i = 1, \ldots, J \qquad (4.7)$$

We must use the incidence matrices, because a row j of an adjacency matrix does only show for what arcs the item with number j is the origin, but it does not provide the information for which arcs the item is a destination.[5]

Two rows, say, j and i (which represent items j and i) are called equivalent if and only if each number that occurs, say, $n \geq 0$ times in row j, also occurs n times in row i and $llc_j = llc_i$. More formally, two rows j and i are equivalent if and only if there is a permutation π of column indices so that

$$llc_j = llc_i \text{ and } \hat{a}_{jk} = \hat{a}_{i\pi(k)} \qquad k = 1, \ldots, J \qquad (4.8)$$

holds.

The procedure to be employed iteratively finds a row j in the one incidence matrix that is equivalent to a row i in the other and deletes both. It terminates if no further equivalent (and undeleted) rows can be found. If all rows are deleted, the two incidence matrices and so the two adjacency matrices do represent isomorphic gozinto–structures.

[5]It is easy to find examples which show that using the adjacency matrices instead of the incidence matrices would let the procedure to be presented indicate an isomorphism for some gozinto–structures which are not identical.

As an example, let us reconsider the gozinto–structures in Figures 4.3 and 4.4 again to see if the presented procedure finds out that both structures are isomorphic. The incidence matrix that corresponds to Figure 4.3 is given at the left hand side and the one that corresponds to Figure 4.4 is given on the right hand side where horizontal lines separate different low levels:

$$\left(\begin{array}{cccccc} 0 & -1 & -1 & -1 & 0 & 0 \\ \hline 1 & 0 & 0 & 0 & -1 & 0 \\ 1 & 0 & 0 & 0 & -2 & -2 \\ 1 & 0 & 0 & 0 & -1 & -1 \\ \hline 0 & 1 & 2 & 1 & 0 & 0 \\ 0 & 0 & 2 & 1 & 0 & 0 \end{array}\right) \stackrel{?}{\Longleftrightarrow} \left(\begin{array}{cccccc} 0 & -1 & -1 & -1 & 0 & 0 \\ \hline 1 & 0 & 0 & 0 & -2 & -2 \\ 1 & 0 & 0 & 0 & -1 & -1 \\ 1 & 0 & 0 & 0 & 0 & -1 \\ \hline 0 & 2 & 1 & 0 & 0 & 0 \\ 0 & 2 & 1 & 1 & 0 & 0 \end{array}\right)$$

Looking for corresponding rows, we find that row 3 of the matrix on the left hand side is equivalent to row 2 of the right hand side matrix. Hence, these rows can be deleted. So can rows 5 and 6, respectively.

$$\left(\begin{array}{cccccc} 0 & -1 & -1 & -1 & 0 & 0 \\ \hline 1 & 0 & 0 & 0 & -1 & 0 \\ \times & \times & \times & \times & \times & \times \\ 1 & 0 & 0 & 0 & -1 & -1 \\ \hline \times & \times & \times & \times & \times & \times \\ 0 & 0 & 2 & 1 & 0 & 0 \end{array}\right) \stackrel{?}{\Longleftrightarrow} \left(\begin{array}{cccccc} 0 & -1 & -1 & -1 & 0 & 0 \\ \hline \times & \times & \times & \times & \times & \times \\ 1 & 0 & 0 & 0 & -1 & -1 \\ 1 & 0 & 0 & 0 & 0 & -1 \\ \hline 0 & 2 & 1 & 0 & 0 & 0 \\ \times & \times & \times & \times & \times & \times \end{array}\right)$$

The reader may convince himself that moving on indeed deletes all rows and thus gives the desired result that both matrices represent isomorphic gozinto–structures.

4.1.2 Generation of External Demand

The parameters for generating external demand matrices are given in Table 4.5.

These parameters guide the random generation of demand matrices. While generating a d_{jt}–value for $j = 1, \ldots, J$ and $t = 1, \ldots, T$, three cases may occur:

Case 1: $t \leq T_{idle} T_{micro}$ or $\frac{t}{T_{micro}} \neq \lfloor \frac{t}{T_{micro}} \rfloor$

Case 2: $t > T_{idle} T_{micro}$ and $\frac{t}{T_{micro}} = \lfloor \frac{t}{T_{micro}} \rfloor$ and $\mathcal{S}_j \neq \emptyset$ and $TYPE_D = \textit{end items}$

Symbol	Definition
$[DEM_{min}, DEM_{max}]$	Interval of demand sizes where $DEM_{max} \geq DEM_{min} \geq 0$.
$ITEMS(0)$	Number of end items.
J	Total number of items where $J \geq ITEMS(0)$.
N_D	Number of different external demand matrices to be created.
T_{idle}	Number of idle macro periods where $T_{idle} \geq 0$.
T_{macro}	Number of macro periods where $T_{macro} > T_{idle}$.
T_{micro}	Number if micro periods where $T_{micro} > 0$.
$TYPE_D$	Indicator for which items external demand may occur, i.e. $TYPE_D \in \{end\ items, all\ items\}$.

Table 4.5: Parameters for Generating External Demand

Case 3: $t > T_{idle}T_{micro}$ and $\frac{t}{T_{micro}} = \lfloor \frac{t}{T_{micro}} \rfloor$ and ($\mathcal{S}_j = \emptyset$ or $TYPE_D = all\ items$)

Depending on the case that holds, we define

$$d_{jt} \stackrel{def}{=} \begin{cases} 0 & Case1 \\ 0 & Case2 \\ IRAND[DEM_{min}, DEM_{max}] & Case3 \end{cases} \tag{4.9}$$

where $T = T_{macro}T_{micro}$. The idea behind this definition of T links model specific and real–world points of view. In the real–world external demands are not to be met at arbitrary points of time, but at some well–defined points of time such as the end of a shift, the end of a day, or the end of a week, for example. These real–world macro periods may then be subdivided into fine grain micro periods which are the subject of consideration when solving an instance. Since the user may set $T_{micro} = 1$ and hence $T = T_{macro}$

this way of generating external demand is not restrictive, but close to real–world.[6] The parameter T_{idle} specifies a number of macro periods at the beginning of the planning horizon in which no external demand occurs. This allows to generate multi–level instances with low initial inventory where the depth of items must fit into the time window between period one and the due date of orders.

In total a number of N_D different external demand matrices is generated where checking if two matrices are equal can be done with a straightforward comparison.

4.1.3 Generation of Capacity Limits

To generate a matrix of capacity limits we need a gozinto–structure, an external demand matrix which defines external demand for each item in the gozinto–structure, v_j–, p_j–, and m_j–vectors, and a parameter U where $0 < U \leq 100$ which defines the percentage of capacity utilization per machine[7] as input.

Basically, the capacity utilization is a measure which relates capacity demand to capacity availability. Before we give more details, let us first introduce an auxiliary notation for what (external or internal) demand is to be met at what point of time. Disregarding restrictive assumptions about how many items can be produced per period, a lot–for–lot policy leads to

$$d_{jt}^{L4L} \stackrel{def}{=} \begin{cases} d_{jt} + \sum_{i \in S_j} a_{ji} d_{i(t+v_j)}^{L4L} & , \text{if } 1 \leq t \leq T - v_j \\ d_{jt} & , \text{if } T - v_j < t \leq T \end{cases} \qquad \begin{array}{l} j = 1, \ldots, J \\ t = 1, \ldots, T \end{array} \qquad (4.10)$$

where, by the way, $d_{jt}^{L4L} = d_{jt}$ for all $j = 1, \ldots, J$ and $t = 1, \ldots, T$ in the single–level case.

Most authors define the capacity utilization of a machine m

[6] See also [Fle90, Fle94] for a discussion of macro and micro periods when interpreting the CLSP and the DLSP.

[7] A value $U > 100$ would indicate that capacity needs exceed the capacity availability. As we do not consider overtime such values are not valid.

where $m = 1, \ldots, M$ as:[8]

$$U_m \stackrel{def}{=} \frac{\sum_{j \in \mathcal{J}_m} \sum_{t=1}^{T} p_j d_{jt}^{L4L}}{\sum_{t=1}^{T} C_{mt}} 100 \tag{4.11}$$

So, if we assume $U_1 \approx \ldots \approx U_M \approx U$ it seems to be a good idea to choose

$$C_{mt} = \left\lceil \frac{\sum_{j \in \mathcal{J}_m} \sum_{t=1}^{T} p_j d_{jt}^{L4L}}{T \cdot U} 100 \right\rceil . \tag{4.12}$$

But, since demand is dynamic this is not a good choice as a small single–item example shows. Suppose, $J = 1$, $M = 1$, and $T = 4$. Furthermore, assume the data given in Table 4.6. Let $U = 70$.

d_{jt}	$t = 1$	2	3	4	p_j
$j = 1$		35		10	1

Table 4.6: A Small Single–Item Example for Determining the Capacity Utilization

Using formula (4.12) gives $C_{1t} = 17$ for all $t = 1, \ldots, T$. Unfortunately, there is no feasible solution for an instance with these parameters, because meeting the demand in period 2 exceeds the available capacity.

The problem here is that using an overall average as defined by (4.12) does not take the variance of demand into account. A procedure which reflects the dynamic nature of demands is thus

[8] A somehow more realistic definition would be

$$U_m \stackrel{def}{=} \frac{\sum_{j \in \mathcal{J}_m} p_j nr_j}{\sum_{t=1}^{T} C_{mt}} 100,$$

because in the presence of positive initial inventory the capacity utilization is overestimated, otherwise.

needed to generate a capacity limit matrix. The method that we use is given in Table 4.7 where

$$\overline{d}_{m\hat{t}t} \stackrel{def}{=} \frac{\sum_{j\in\mathcal{J}_m}\sum_{\tau=\hat{t}}^{t} p_j d_{j\tau}^{L4L}}{t-\hat{t}+1} \qquad 1 \leq \hat{t} \leq t \leq T \tag{4.13}$$

is used to denote the average capacity demand in the interval $[\hat{t},\ldots,t]$ of periods.

---

for $m = 1$ to $m = M$
    $t := 1.$
    while $t \leq T$
        $\hat{t} := t.$
        $\tilde{d}_m := 0.$
        while $\tilde{d}_m \leq \overline{d}_{m\hat{t}t}$ and $t \leq T$
            $\tilde{d}_m := \overline{d}_{m\hat{t}t}.$
            $t := t+1.$
        for $\tau = \hat{t}$ to $\tau = t-1$
            $C_{m\tau} := \lceil \tilde{d}_m \frac{100}{U} \rceil.$

Table 4.7: Method to Compute Capacity Limits

The presented method guarantees that

$$\sum_{j\in\mathcal{J}_m}\sum_{\tau=1}^{t} p_j d_{j\tau}^{L4L} \leq \frac{U}{100}\sum_{\tau=1}^{t} C_{m\tau} \qquad \begin{array}{l} m=1,\ldots,M \\ t=1,\ldots,T \end{array} \tag{4.14}$$

holds. Also,

$$\frac{\sum_{j\in\mathcal{J}_m}\sum_{t=1}^{T} d_{jt}^{L4L}}{T\cdot\frac{U}{100}+\sum_{j\in\mathcal{J}_m}\sum_{t=1}^{T} d_{jt}^{L4L}}U \leq U_m \leq U \qquad m=1,\ldots,M \tag{4.15}$$

is a valid worst case bound which proves a satisfying approximation for $U_1 \approx \ldots \approx U_M \approx U$,[9] because if

$$\sum_{j \in \mathcal{J}_m} \sum_{t=1}^{T} d_{jt}^{L4L} >> T$$

which certainly is true in most real–world situations then the left hand side evaluates to

$$\frac{\sum_{j \in \mathcal{J}_m} \sum_{t=1}^{T} d_{jt}^{L4L}}{T \cdot \frac{U}{100} + \sum_{j \in \mathcal{J}_m} \sum_{t=1}^{T} d_{jt}^{L4L}} U \approx U. \tag{4.16}$$

As an example, suppose the gozinto–structure given in Figure 4.5 where $J = 5$. Furthermore, assume $M = 2$ and $T = 10$. All other relevant parameters are provided in Table 4.8 where external demand occurs for item 1 only. Table 4.9 shows a protocol of running the procedure given in Table 4.7 with these data and $U = 70$.

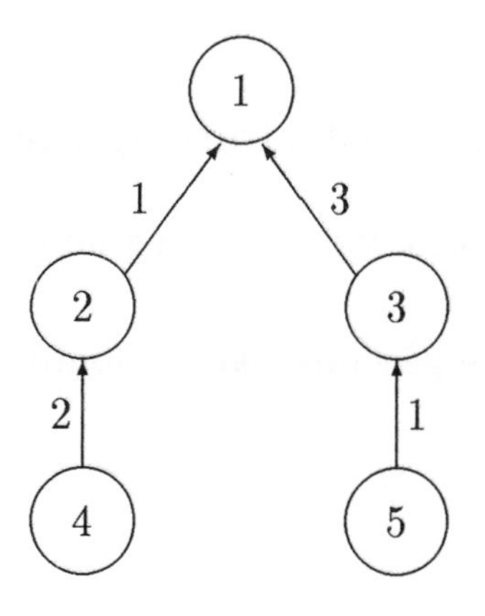

Figure 4.5: A Gozinto–Structure with Five Items

[9] Note, if we would have chosen to compute $C_{mT} := \tilde{d}_m \frac{100}{U}$ in the method given in Table 4.7 then we would have $U_1 = \ldots = U_M = U$. However, we decided to generate integer values for demand (see (4.9)) as well as for capacity limits.

d_{jt}^{L4L}	$t=1$	2	3	4	5	6	7	8	9	10	m_j	p_j	v_j
$j=1$					10					20	1	1	1
$j=2$				10					20		2	1	1
$j=3$				30					60		1	1	1
$j=4$			20					40			2	1	1
$j=5$			30					60			1	1	1

Table 4.8: Relevant Data for the Example

The entries in the columns $\tilde{d}_1$ and $\tilde{d}_2$, respectively, which are highlighted with boxes trigger the computation of $C_{1\tau}$ and $C_{2\tau}$, respectively, for $\hat{t} \leq \tau \leq t$. For example, since $\tilde{d}_2 = 12$ is a local maximum when computing average capacity demands beginning with period 5 stepwise up to period 9, $C_{25} = C_{26} = C_{27} = C_{28} = C_{29} = \lceil 12\frac{100}{70} \rceil = 18$. Due to (4.11), $U_1 = 69.08$ and $U_2 = 67.16$ which is close to the desired value $U = 70$.

4.1.4 Generation of Holding and Setup Costs

The parameters for generating holding and setup costs are given in Table 4.10.

Holding costs are generated at random using[10]

$$h_j \in IRAND[\ \left\lfloor HCOST_{min} + (J-j)\frac{HCOST_{max} - HCOST_{min}}{J} \right\rfloor, \left\lceil HCOST_{min} + (J-j+1)\frac{HCOST_{max} - HCOST_{min}}{J} \right\rceil] \quad (4.17)$$

for $j = 1, \ldots, J$. The generation of holding costs therefore tends to assign high holding costs to items with a small low level code and low holding costs to items with a large low level code (see Subsection 4.1.1).

[10] Using the $RRAND$-function would be fine as well.

$\hat{t}$	t	$\overline{d}_{1\hat{t}t}$	$\tilde{d}_1$	$\overline{d}_{2\hat{t}t}$	$\tilde{d}_2$	C_{1t}	C_{2t}
1			0		0		
	1	0	0	0	0	22	11
	2	0	0	0	0	22	11
	3	10	10	6.67	6.67	22	11
	4	15	15	7.5	7.5	22	11
	5	14		6			
5			0		0		
	5	10	10	0	0	15	18
	6	5		0	0		18
6			0				
	6	0	0			43	
	7	0	0	0	0	43	18
	8	20	20	10	10	43	18
	9	30	30	12	12	43	18
	10	28		10			
10			0		0		
	10	20	20	0	0	29	0
	11						

Table 4.9: A Protocol of the Method in Table 4.7

Setup costs can now be chosen with respect to[11]

$$s_j \in IRAND[\ \left\lfloor COSTRATIO \cdot h_j(1 - \frac{RATIODEV}{100})\right\rfloor, \quad (4.18)$$

$$\left\lceil COSTRATIO \cdot h_j(1 + \frac{RATIODEV}{100})\right\rceil]$$

for $j = 1, \ldots, J$.

To find N_C different cost vectors a simple comparison helps to check for doubles.

[11]Again, the $RRAND$–function would be okay, too.

Symbol	Definition
$COSTRATIO$	Ratio of setup and holding costs where $COSTRATIO \geq 0$.
$[HCOST_{min}, HCOST_{max}]$	Interval of holding costs where $HCOST_{max} \geq HCOST_{min} \geq 0$.
J	Number of items.
N_C	Number of different cost vectors to be created.
$RATIODEV$	Acceptable deviation from the cost ratio where $0 \leq RATIODEV \leq 100$.

Table 4.10: Parameters for Generating Holding and Setup Costs

4.1.5 Generating PLSP–PM Instances

The generation of PLSP–PM instances is a more or less straightforward extension of the generation of PLSP–MM instances as values for s_{jm} and y_{jm0} for $j = 1, \ldots, J$ and $m = 1, \ldots, M$ are concerned. The values p_{jm} for $j = 1, \ldots, J$ and $m = 1, \ldots, M$ must fulfill $\mathcal{J}_m \neq \emptyset$ for $m = 1, \ldots, M$ and $| \mathcal{M}_j |= MACHPERITEM$ for $j = 1, \ldots, J$ with $MACHPERITEM > 0$ being a user specified parameter. The generation of such values needs no further explanation.

The method given in Table 4.7 which determines capacity limits stays the same, but, the definition of $\overline{d}_{m\hat{t}t}$ slightly changes to

$$\overline{d}_{m\hat{t}t} \stackrel{def}{=} \frac{\sum_{j \in \mathcal{J}_m} \sum_{\tau=\hat{t}}^{t} \frac{p_{jm} d_{j\tau}^{L4L}}{MACHPERITEM}}{t - \hat{t} + 1} \qquad 1 \leq \hat{t} \leq t \leq T \quad (4.19)$$

for $m = 1, \ldots, M$ to take into account that several machines may be used for production.

4.1.6 Generating PLSP–MR Instances

Analogously to the PLSP–PM instance generation we compute p_{jm}, s_{jm}, and y_{jm0} for $j = 1, \ldots, J$ and $m = 1, \ldots, M$ for PLSP–MR instances.

To determine capacity limits with the method given in Table 4.7, we use

$$\overline{d}_{m\hat{t}t} \stackrel{def}{=} \frac{\sum_{j \in \mathcal{J}_m} \sum_{\tau=\hat{t}}^{t} p_{jm} d_{j\tau}^{L4L}}{t - \hat{t} + 1} \qquad 1 \leq \hat{t} \leq t \leq T \qquad (4.20)$$

for $m = 1, \ldots, M$.

4.1.7 Generating PLSP–PRR Instances

The generation of PLSP–PRR instances closely follows the lines of the PLSP–MM instance generation. $\tilde{M}$, $\tilde{L}$ and $\tilde{T}$ are user specified where $\tilde{L} \cdot \tilde{T} = T$ must hold. For $j = 1, \ldots, J$ and $m = 1, \ldots, \tilde{M}$ we choose $\tilde{p}_{jm} \in RRAND[CAPNEED_{min}, CAPNEED_{max}]$. The capacity limits $\tilde{C}_{mt}$ for $m = 1, \ldots, \tilde{M}$ and $t = 1, \ldots, \tilde{T}$ are determined using a straightforward modification of the procedure given in Table 4.7 where $\tilde{M}$ and $\tilde{T}$, are used instead of M and T, respectively. Let us use

$$\overline{d}_{m\hat{t}t} \stackrel{def}{=} \frac{\sum_{j \in \tilde{\mathcal{J}}_m} \sum_{\tau=(\hat{t}-1)\tilde{L}+1}^{t\tilde{L}} \tilde{p}_{jm} d_{j\tau}^{L4L}}{t - \hat{t} + 1} \qquad 1 \leq \hat{t} \leq t \leq \tilde{T} \quad (4.21)$$

for $m = 1, \ldots, \tilde{M}$ to compute the $\tilde{C}_{mt}$–matrix. Later, we will use $\tilde{U}$ to refer to the capacity utilization of the partially renewable resources and U to refer to the capacity utilization of the machine for which a setup state is considered.

4.1.8 Some Remarks

Note that the instance generator does not guarantee the existence of a feasible solution for any instance that can be generated.

Our implementation of the instance generator APCIG described in this section produces two types of output. First, it creates

C programming language header files which can than be included in C–coded programs and thus allows convenient access to the parameters. Second, it generates LINGO data files which makes using the LINGO standard solver [LINDO93] easy to employ either for solving a small instance optimally or to generate so–called MPS[12]–formatted data. Since both types of output are in ASCII[13] text other self–made converters will have easy access.

4.2 Experimental Design for the PLSP–MM

Both, lower bound as well as upper bound procedures for the PLSP–MM will be tested on small instances which can be solved optimally with the standard MIP–solver LINGO [LINDO93]. To avoid describing the same test–bed over and over again at different positions in the book, we define the test instances here once and for all. Later on, we will refer to these as small PLSP–MM instances.

Due to preliminary computational studies, we know what parameters have almost no influence on the outcome of computational studies, what parameters have low influence, and what parameters highly influence the outcome of a computational study. According to this insight we let some parameters constant in all tests, using common random numbers for some others, and systematically varied the third.

All instances are generated using APCIG as described above.

4.2.1 Constant Parameters

Whether or not a PLSP–MM instance can be solved optimally with standard solvers within reasonable time basically depends on the number of binary variables and hence on the choice of J and T. To define small test instances we choose $J = 5$ and $T = 10$

[12]Mathematical Programming System [HuRoWa94].

[13]American Standard Code for Information Interchange.

which is small enough to be solved optimally and large enough to construct non–trivial, multi–level instances. The average run–time for solving instances of that size with LINGO optimally is about two to three hours per instance. These measures are gained without using any of the valid inequalities given in Subsection 3.3.7.

Furthermore, we choose the following parameters:

Initial inventory:
$INITINV_{min} = INITINV_{max} = 0$
and thus $I_{j0} = 0$ for $j = 1, \ldots, J$.

Capacity requirements:
$CAPNEED_{min} = CAPNEED_{max} = 1$
and thus $p_j = 1$ for $j = 1, \ldots, J$.

Lead time:
$LEADTIME_{min} = LEADTIME_{max} = 1$
and thus $v_j = 1$ for $j = 1, \ldots, J$.

Initial setup state:
$y_{j0} = 0$ for $j = 1, \ldots, J$.

Gozinto–structures:
$ARC_{min} = 1$.
$ARC_{max} = 3$.
$DEPTH = 2$.
$ITEMS(0) = 2$.
$PRIVAL(1) = PRIVAL(2) = 1$.
$TYPE_G = general$.

External demand:
$DEM_{min} = 40$.
$DEM_{max} = 50$.
$TYPE_D = end\ items$.

Holding and setup costs:
$HCOST_{min} = 1.$
$HCOST_{max} = 10.$
$RATIODEV = 10.$

4.2.2 Systematically Varied Parameters

For those parameters which are assumed to have a significant impact on the computational outcome, we choose a full factorial design. The parameter levels are:

Number of machines:
$M \in \{1, 2\}.$

Complexity of the gozinto–structures:
$COMPLEXITY \in \{0.2, 0.8\}.$

Demand patterns:
$(T_{macro}, T_{micro}, T_{idle}) \in \{(10, 1, 5), (5, 2, 2), (1, 10, 0)\}.$

Capacity utilization:
$U \in \{30, 50, 70\}.$

Ratio of setup and holding costs:
$COSTRATIO \in \{5, 150, 900\}.$

In total we thus have $2 \cdot 2 \cdot 3 \cdot 3 \cdot 3 = 108$ parameter level combinations.

4.2.3 Randomly Generated Parameters

For each parameter combination of the systematically varied parameters we generated 10 replications using common random numbers, i.e. $N_G = N_D = N_C = 10$. This gives a total of $10 \cdot 108 = 1,080$ small test instances for the PLSP–MM. It turned out that 47 of these instances have no feasible solution. Hence, we have a test–bed of 1,033 instances for which a feasible solution exists.

4.3 A Comparison of Computing Machines

For the computational tests conducted in this book, we will consider the run–time performance of the methods under concern. These results do heavily depend on the computer being used. Hence, we provide a run–time performance index for a comparison between different platforms in Table 4.11 where all run–times are given in CPU–seconds. For doing so, 20 PLSP–MM instances are arbitrarily chosen and solved with one of the heuristics to be presented in subsequent chapters. The details of the 20 instances as well as an understanding of how the heuristic works is not important at this point. All run–times given later on are measured on a Pentium P120 computer running a LINUX operating system.

Platform:	486DX 25 MHz	Pentium 75 MHz	Power PC 601 80 MHz	Pentium 120 MHz
1	3.24	0.29	0.24	0.11
2	4.18	0.42	0.36	0.16
3	7.69	0.62	0.47	0.23
4	9.62	0.86	0.65	0.30
5	9.89	0.92	0.68	0.31
6	10.33	0.95	0.72	0.33
7	13.74	1.49	1.10	0.50
8	15.05	1.60	1.25	0.55
9	15.49	1.83	1.30	0.59
10	16.70	1.99	1.44	0.62
11	17.25	1.99	1.52	0.66
12	19.67	2.34	1.79	0.74
13	19.78	2.38	1.82	0.77
14	16.43	2.05	1.46	0.64
15	15.44	1.96	1.35	0.60
16	18.35	2.46	1.71	0.75
17	20.22	2.78	2.02	0.88
18	33.30	4.69	3.58	1.50
19	28.68	4.09	3.05	1.28
20	34.34	5.27	3.83	1.82
Total:	329.39	40.98	30.34	13.34
Index:	24.69	3.07	2.27	1.00

Table 4.11: Run–Time Performance Indices

Chapter 5

Lower Bounds

Where standard MIP–solvers fail to compute optimum objective function values for PLSP–instances, lower bounds may be used as a point of reference for evaluation purposes. In this chapter, we compute lower bounds for the PLSP–MM. Solving the LP–relaxation of a PLSP–MM–model optimally is a straightforward idea. Section 5.1 deals with this approach and discusses a network reformulation of the model. Another way to get lower bounds is to ignore some of the constraints and to solve the remaining problem optimally. This path is followed in Section 5.2 where a B&B–procedure is used to attack the uncapacitated, multi–level, multi–machine lot sizing and scheduling problem.[1] On the basis of this, Section 5.3 introduces a method to solve a Lagrangean relaxation of the capacity constraints. Finally, Section 5.4 summarizes the lower bounds obtained.

5.1 Network Representations

Some researchers have introduced network representations for lot sizing problems to derive "tighter" reformulations, i.e. new models

[1]Note, solving an uncapacitated lot sizing and scheduling problem is far more than just finding a value less than or equal to the optimum objective function value of a PLSP–MM–instance, or lower bound for short. It also provides a solution for the uncapacitated problem which may appear in distribution networks and supply chains for instance [LeBi93, LeBiCa93, SiEr94a].

where the LP–relaxation of an instance has an optimum objective function value that is greater than the optimum objective function value of the LP–relaxations of the straightforward model formulation [EpMa87, Ros86, Sta94, Sta95].

5.1.1 A Simple Plant Location Representation

For the multi–level CLSP, a computational study in [Sta94] reveals that a simple plant location representation adapted from [Ros86] gives the same lower bounds as a shortest route representation adapted from [EpMa87].[2]

In here, we thus confine our focus of attention to a simple plant location representation of the PLSP–MM. Table 5.1 gives a new decision variable where the computation of the gross demand is defined in (4.10). All other notation is as defined in Chapter 3.

Symbol	Definition
$z_{jt\tau}$	Fraction of the gross demand $d_{j\tau}^{L4L}$ for item j produced in period t where $z_{jt\tau} \in [0,1]$.

Table 5.1: A New Decision Variable for the PLSP–MM Network Representation

$$\min \sum_{j=1}^{J} \sum_{t=1}^{T} (s_j x_{jt} + h_j I_{jt}) \tag{5.1}$$

subject to

$$I_{jt} = I_{j(t-1)} + \sum_{\tau=t}^{T} z_{jt\tau} d_{j\tau}^{L4L} - d_{jt} - \sum_{i \in \mathcal{S}_j} a_{ji} \sum_{\tau=t}^{T} z_{it\tau} d_{i\tau}^{L4L} \qquad \begin{array}{l} j = 1, \ldots, J \\ t = 1, \ldots, T \end{array} \tag{5.2}$$

[2]For a theory of variable redefinition as used in [EpMa87] see [Mar87].

$$I_{jt} \geq \sum_{i \in \mathcal{S}_j} \sum_{\tau=t+1}^{\min\{t+v_j,T\}} a_{ji} \sum_{\hat{\tau}=\tau}^{T} z_{i\tau\hat{\tau}} d_{i\hat{\tau}}^{L4L} \qquad \begin{array}{l} j = 1,\ldots,J \\ t = 0,\ldots,T-1 \end{array} \qquad (5.3)$$

$$\sum_{j \in \mathcal{J}_m} y_{jt} \leq 1 \qquad \begin{array}{l} m = 1,\ldots,M \\ t = 1,\ldots,T \end{array} \qquad (5.4)$$

$$x_{jt} \geq y_{jt} - y_{j(t-1)} \qquad \begin{array}{l} j = 1,\ldots,J \\ t = 1,\ldots,T \end{array} \qquad (5.5)$$

$$z_{jt\tau} \leq y_{j(t-1)} + y_{jt} \qquad \begin{array}{l} j = 1,\ldots,J \\ t = 1,\ldots,T \\ \tau = t,\ldots,T \end{array} \qquad (5.6)$$

$$\sum_{j \in \mathcal{J}_m} p_j \sum_{\tau=t}^{T} z_{jt\tau} d_{j\tau}^{L4L} \leq C_{mt} \qquad \begin{array}{l} m = 1,\ldots,M \\ t = 1,\ldots,T \end{array} \qquad (5.7)$$

$$\sum_{\tau=1}^{t} z_{j\tau t} \leq 1 \qquad \begin{array}{l} j = 1,\ldots,J \\ t = 1,\ldots,T \end{array} \qquad (5.8)$$

$$y_{jt} \in \{0,1\} \qquad \begin{array}{l} j = 1,\ldots,J \\ t = 1,\ldots,T \end{array} \qquad (5.9)$$

$$I_{jt}, x_{jt} \geq 0 \qquad \begin{array}{l} j = 1,\ldots,J \\ t = 1,\ldots,T \end{array} \qquad (5.10)$$

$$z_{jt\tau} \geq 0 \qquad \begin{array}{l} j = 1,\ldots,J \\ t = 1,\ldots,T \\ \tau = t,\ldots,T \end{array} \qquad (5.11)$$

To get this model formulation we simply replaced the decision variables q_{jt} in the PLSP–MM model given in Chapter 3 with $\sum_{\tau=t}^{T} z_{jt\tau} d_{j\tau}^{L4L}$. Furthermore, we added the constraints (5.8) which reflect the fact that the new decision variable represents a fraction of gross demand.[3] This, by the way, allows us to simplify the constraints (3.6) now stated as (5.6).

[3]Noteworthy to say that $\sum_{\tau=1}^{t} z_{j\tau t} = 1$ for $j = 1,\ldots,J$ and $t = 1,\ldots,T$ is not valid in the presence of initial inventory.

5.1.2 Experimental Evaluation

In a computational study we compare the LP–relaxation of the original PLSP–model[4] with the LP–relaxation of the simple plant location model for the PLSP. As a test–bed we use the small PLSP–instances given in Section 4.2. Tables A.1 to A.3 provide the results for the original model, while Tables A.4 to A.6 in the Appendix A show the yieldings of a "tight" reformulation as a simple plant location problem. For each combination of the systematically varied parameters U (the capacity utilization), M (the number of machines), $\mathcal{C}$ (the gozinto–structure complexity), $COSTRATIO$ (the ratio of setup and holding costs), and $(T_{macro}, T_{micro}, T_{idle})$ (the demand pattern) these tables give the average deviation from the optimum objective function value of the PLSP–instance, the so–called integrality gap. For each instance this deviation is computed as

$$deviation \stackrel{def}{=} 100\frac{LB - OPT}{OPT} \tag{5.12}$$

where LB is the lower bound determined by means of the LP–relaxation and OPT is the optimum objective function value of the PLSP–instance. Note, a deviation close to zero is desired.

For each parameter level combination we also give in brackets the worst case deviation and the number of PLSP–instances for which a feasible solution exists. Due to our test–bed, this number is less than or equal to 10 for each parameter level combination. Summing these numbers up gives 1,033, the total number of instances in the test–bed with a feasible solution. As one can see, for each parameter level combination there are at least six instances for which a feasible solution exists.

Consider for example the results for the original PLSP–model where $U = 30$, $M = 1$, $\mathcal{C} = 0.2$, $COSTRATIO = 5$, and $(T_{macro}, T_{micro}, T_{idle}) = (10, 1, 5)$. As we can see, all 10 PLSP–instances have a feasible solution. The average deviation of the lower bound from the optimum result is -41.91%. At least one instance gives a deviation of -55.09% which is the worst case result

[4]None of the valid inequalities given in Subsection 3.3.7 is integrated into the model.

of all 10 instances.

For both model formulations a good performance, say with a deviation in the -10% scope, comes out for instances with $M = 2$, $COSTRATIO = 5$, and $(T_{macro}, T_{micro}, T_{idle}) = (1, 10, 0)$ only. All other parameter level combinations give much worse results. Some of them give a whopping -70% or more which is often the case if $COSTRATIO = 900$.

To ease a detailed analysis of these results we aggregate the data. Table 5.2 reveals the impact of the parameter M and provides the average integrality gap. Apparently, increasing the number of machines reduces the deviation from the optimum results decidedly. This matches a former result given in [Kim94b] where it was shown that, if $M = J$ which is the maximum number of machines in a multi-machine setting, the problem of optimizing the PLSP-MIP-model formulation reduces to be an LP-problem.

	$M = 1$	$M = 2$
PLSP-Model	-55.49	-40.96
Simple Plant Location Model	-47.70	-37.30

Table 5.2: The Impact of the Number of Machines on the Integrality Gap

Whether or not the complexity of the gozinto-structure plays a role can be read in Table 5.3. Though the impact of the complexity is not dramatic, it seems that gozinto-structures with a high complexity give an average gap that is slightly closer for the original PLSP-model formulation. When solving a plant location model, gozinto-structure complexity has almost no effect.

In Table 5.4 we see what happens to the lower bound if the demand pattern changes. While for the original PLSP-model a sparsely-filled demand matrix gives the best result, for the plant location model a demand matrix with many non-zeroes turns out to be advantageous on average. The original PLSP-model seems to be more sensitive to different demand patterns, because the variance of the average results is greater than for the simple plant

	$\mathcal{C} = 0.2$	$\mathcal{C} = 0.8$
PLSP–Model	-49.48	-46.88
Simple Plant Location Model	-42.68	-42.29

Table 5.3: The Impact of the Gozinto–Structure Complexity on the Integrality Gap

	$(T_{macro}, T_{micro}, T_{idle}) =$		
	$(10, 1, 5)$	$(5, 2, 2)$	$(1, 10, 0)$
PLSP–Model	-48.52	-51.72	-44.71
Simple Plant Location Model	-40.47	-44.43	-42.71

Table 5.4: The Impact of the Demand Pattern on the Integrality Gap

location model.

The ratio of setup and holding costs significantly affects the quality of the lower bounds as can be seen in Table 5.5. Low setup costs result in small integrality gaps for both, the original PLSP–model and the simple plant location model. The higher the setup costs are, the greater the gap. The explanation for this phenomenon is that in an LP–relaxation only a small fraction of the actual setup costs are charged. If setup costs are low, the optimum objective function value of a PLSP–instance mainly is a sum of holding costs. If setup costs are large, the objective function value in large parts is a sum of setup costs. Hence, LP–relaxations perform better when setup costs are low. However, even if setup costs would be zero the LP–relaxation would not necessarily give a zero integrality gap. This is, because in a solution of the LP–relaxation the setup state of a machine is no longer uniquely defined, by definition. As a consequence, more than two items may be produced per period which leads to infeasible production plans and to an underestimation of total holding costs.

The capacity utilization is assumed to be of great importance

	$COSTRATIO =$		
	5	150	900
PLSP–Model	-24.72	-50.72	-69.29
Simple Plant Location Model	-18.70	-44.66	-64.23

Table 5.5: The Impact of the Cost Structure on the Integrality Gap

for the performance of lot sizing and scheduling. For computing lower bounds via an LP–relaxation, however, Table 5.6 shows that the capacity utilization is not the reason for dramatic performance differences. For the original PLSP–model varying the capacity utilization gives no significant changes. For the simple plant location model there is a tendency that the quality of the lower bounds decreases if the capacity utilization increases.

	$U = 30$	$U = 50$	$U = 70$
PLSP–Model	-47.97	-48.53	-48.10
Simple Plant Location Model	-40.10	-42.59	-45.04

Table 5.6: The Impact of the Capacity Utilization on the Integrality Gap

Using LINGO to solve the LP–relaxations takes on average less than a minute per instance no matter what model is used.

In summary, we can state that the performance of the simple plant location model formulation slightly outperforms the original PLSP–model formulation. The overall average result for the former is a -42.49% integrality gap, while the latter yields a gap of -48.20% on average. In contrast to lot sizing without scheduling, a plant location reformulation for lot sizing and scheduling does not give a sharp lower bound by solving the LP–relaxation. Hence, we need other ways to determine lower bounds.

5.2 Capacity Relaxation

5.2.1 Motivation for Relaxing Capacity Constraints

Many hard–to–solve problems, and so the PLSP–MM, can be modelled as an easier–to–solve problem complicated by a set of constraints. Removing this set of constraints from the PLSP–model yields a model which defines an optimum objective function value that is a lower bound for the PLSP. So, we should find out which constraints make instances of the PLSP–MM–model hard to solve, remove these constraints, and develop an exact solution procedure for the remaining problem which may then be used to find lower bounds. The fundamental motivation here is that a tailor–made method is much more efficient (primarily in terms of run–time) than a standard MIP–solver.

As we have learned in the preceding section, violating the binary conditions for the setup state variables gives poor results. Thus, more promising approaches should respect integrality. To see what else makes the PLSP–MM be a hard–to–solve problem, suppose that all binary variables are fixed, say, to the values that occur in an optimum solution. What is left then is a linear program still not easy to solve (optimally).[5] For example, as we have discussed in Chapter 3, lot splitting may need to occur which makes the computation of production quantities a non–trivial task. So, the question is which constraints must be dropped to give a problem in which lot splitting does not occur any more. Due to the results given in Subsection 3.3.6 we drop the capacity constraints of the PLSP–MM and assume that initial inventory is zero.[6] The resulting uncapacitated PLSP–MM is denoted as U–PLSP for short.

Under these assumptions optimum production quantities equal

[5] Of course, standard LP–solvers may be used, but, as we shall see we require more efficient procedures.

[6] The latter assumption is restrictive. But, remember that lower bounds will later on be used to test the performance of PLSP–MM–heuristics. As we expect the initial inventory to have no significant impact on the performance of these heuristics, a test–bed with no initial inventory seems to be sufficient (see Section 4.2).

the sum of subsequent order sizes. More formally, in an optimum solution of an U–PLSP–instance we either have $q_{jt} = 0$ or

$$q_{jt} = CD_{jt} \stackrel{def}{=} \sum_{\tau=t}^{\hat{t}} d_{jt} + \sum_{i \in \mathcal{S}_j} \sum_{\tau=t+v_j}^{\min\{\hat{t}+v_j, T\}} a_{ji} q_{i(\tau+v_j)} \tag{5.13}$$

for $j = 1, \ldots, J$, $t = 1, \ldots, T$, and some $\hat{t}$ for which $t \leq \hat{t} \leq T$ holds. CD_{jt} denotes the cumulative future demand for item j in period t not been met by production in periods later than t. Thus, the matrix of production quantities can also be represented by an integer matrix

$$mask_{jt} \stackrel{def}{=} \begin{cases} 0 & \text{, if } q_{jt} = 0 \\ > 0 & \text{, if } q_{jt} > 0 \end{cases} \qquad \begin{array}{l} j = 1, \ldots, J \\ t = 1, \ldots, T \end{array} \tag{5.14}$$

to which we will refer to as a production mask.

On the other hand, given a $mask_{jt}$–matrix the q_{jt}–matrix can uniquely be restored using the following rule for each pair of indices $j = 1, \ldots, J$ and $t = 1, \ldots, T$: If $mask_{jt} = 0$ then $q_{jt} = 0$. If $mask_{jt} > 0$ then q_{jt} is computed using formula (5.13) with

$$\hat{t} = \min(\{\tau \mid t \leq \tau < T \wedge mask_{j(\tau+1)} > 0\} \cup \{T\}). \tag{5.15}$$

Bringing things together, to solve an U–PLSP–instance optimally, we may enumerate all y_{jt}–matrices where (3.4) must hold. Given a y_{jt}–matrix, a x_{jt}–matrix that causes minimum setup costs can easily be derived by setting $x_{jt} = 1$ if $y_{j(t-1)} = 0$ and $y_{jt} = 1$ to fulfill (3.5). Otherwise, $x_{jt} = 0$. Due to the results given in Subsection 3.3.5, optimum schedules need not be semi–active. Hence, for each setup state matrix we must then enumerate all $mask_{jt}$–matrices where

$$mask_{jt} \leq y_{j(t-1)} + y_{jt} \leq 2 \qquad \begin{array}{l} j = 1, \ldots, J \\ t = 1, \ldots, T \end{array} \tag{5.16}$$

must be valid because of (3.6). Having both, a setup state matrix and a production mask (which is a representation of the production quantities), we must test for feasibility using formulae (3.2), (3.3) and $I_{jt} > 0$ for $j = 1, \ldots, J$ and $t = 1, \ldots, T$.

In summary, it turns out that solving an U–PLSP–instance essentially reduces to enumerating integer matrices which, when guided by problem specific insight, promises to be much more efficient than using standard MIP–solvers.

5.2.2 Basic Enumeration Scheme

The basic working principle for enumerating setup states and production masks is a backward oriented depth–first search moving on from period T to period 1. Let period t be the current focus of attention. We first choose the setup state for each machine where $j_{mt} \in \mathcal{J}_m$ is the item machine m is set up for at the end of period t. Then, we decide for $mask_{j_{mt}(t+1)}$ indicating whether or not item j_{mt} is produced in period $t+1$ which would be allowed, because the setup state at the beginning of period $t+1$ (equal to the setup state at the end of period t) is properly set. Once this is done, $mask_{j_{mt}t}$ is set. Next, period $t-1$ is concerned following the same lines.

More formally, a recursive procedure $uplsp(t, \Delta t, m, p)$ defines the details. Four parameters are passed to this method: $t \in \{1, \ldots, T\}$ the period under concern, $\Delta t \in \{0, 1\}$ indicating whether the setup state in period t should ($\Delta t = 1$) or should not ($\Delta t = 0$) be set, $m \in \{1, \ldots, M\}$ the machine under concern, and $p \in \{0, 1\}$ the value to be used for computing a production mask entry. The idea is to consider the production mask in period $t + \Delta t$. The initial call is $uplsp(T, 1, 1, 1)$ where all $mask_{jt}$– and y_{jt}–values are initialized with zero for $j = 1, \ldots, J$ and $t = 1, \ldots, T$. How to evaluate a call of the form $uplsp(T, 1, \cdot, \cdot)$ is defined in Table 5.7.

If $uplsp(T, 1, \cdot, \cdot)$ is called, the parameter p is of no relevance, because choosing a production mask for period $T+1$ does not make sense. It is important to understand that once we return from a recursive call to the $uplsp$–procedure, the calling procedure loops back choosing another setup state and starting all over again until all setup states are enumerated. Moving stepwise from $m = 1$ to $m = M$, we assign a setup state to every machine at the end of period T. Afterwards, a call to $uplsp(T, 0, 1, 1)$ is made to decide for the production mask in period T. Table 5.8 gives more details.

$ITEMSET_{mT} := \mathcal{J}_m$.
while ($ITEMSET_{mT} \neq \emptyset$)
  choose $j_{mT} \in ITEMSET_{mT}$.
  $ITEMSET_{mT} := ITEMSET_{mT} \backslash \{j_{mT}\}$.
  $y_{j_{mT}T} := 1$.
  if ($m = M$)
    $uplsp(T, 0, 1, 1)$.
  else
    $uplsp(T, 1, m+1, 1)$.
  $y_{j_{mT}T} := 0$.

Table 5.7: Evaluating $uplsp(T, 1, \cdot, \cdot)$

Note, evaluating $uplsp(t, 0, \cdot, \cdot)$ does not require to choose a setup state. The recursive call to $uplsp(t, 0, m, 0)$ implements the enumeration for the values of the parameter p. What is new in this scheme is the call to $uplsp(t-1, 1, \cdot, \cdot)$ to enumerate the setup states at the end of period $t-1$. Table 5.9 provides an implementation of its evaluation.

For the special case $t = 0$ the evaluation of $uplsp(t, 1, \cdot, \cdot)$ is given in Table 5.10. The difference to what is given in Table 5.9 is that for $t = 0$ we have no choice for the setup state, because y_{j0} is given as a parameter for $j = 1, \ldots, J$. Let j_{m0} denote the unique item machine m is initially set up for (assume $j_{m0} = 0$ if machine m is initially setup for no item).

A call to $uplsp(0, 0, \cdot, \cdot)$ indicates a terminal node in which a setup state matrix and a production mask matrix are completely defined. If such a node is reached, we simply have to check feasibility and evaluate the corresponding production plan as described in Subsection 5.2.1. If this plan is feasible and it improves the current best plan, we memorize it. After a complete enumeration we thus have found an optimum solution for an U–PLSP–instance.

$mask_{j_{mt}t} := mask_{j_{mt}t} + p$.
if $(m = M)$
    $uplsp(t-1, 1, 1, 1)$.
else
    $uplsp(t, 0, m+1, 1)$.
$mask_{j_{mt}t} := mask_{j_{mt}t} - p$.
if $(p = 1)$
    $uplsp(t, 0, m, 0)$.

Table 5.8: Evaluating $uplsp(t, 0, \cdot, \cdot)$ where $1 \leq t \leq T$

5.2.3 Branching Rules

We perform a depth–first search. So, a degree of freedom that remains for branching is the sequence in which setup states are enumerated which is represented by the line

$$\text{choose } j_{mt} \in ITEMSET_{mt}$$

in Tables 5.7 and 5.9 where $1 \leq t \leq T$.

Using some priority rule which assigns a priority value $priority_{jt}$ to each item $j \in ITEMSET_{mt}$, items may be chosen in decreasing order with respect to their priority values (ties might be broken with respect to the item index for example).

Two different kinds of priority rules are worth to be discriminated. On the one hand, we may use static rules which depend on the item indices and/or period indices only. On the other hand, we may use dynamic rules which depend on the history of the execution as well. The advantage of the former ones is that items may be sorted before the enumeration starts while the latter ones cause additional overhead for sorting items over and over again whenever the code represented by Tables 5.7 and 5.9 is called.

Some examples for static priority rules are the item index itself

$$priority_{jt} \stackrel{def}{=} \frac{1}{j}, \tag{5.17}$$

$ITEMSET_{mt} := \mathcal{J}_m$.
while ($ITEMSET_{mt} \neq \emptyset$)
    choose $j_{mt} \in ITEMSET_{mt}$.
    $ITEMSET_{mt} := ITEMSET_{mt} \backslash \{j_{mt}\}$.
    $y_{j_{mt}t} := 1$.
    $mask_{j_{mt}(t+1)} := mask_{j_{mt}(t+1)} + p$.
    if ($m = M$)
        $uplsp(t, 0, 1, 1)$.
    else
        $uplsp(t, 1, m + 1, 1)$.
    $mask_{j_{mt}(t+1)} := mask_{j_{mt}(t+1)} - p$.
    $y_{j_{mt}t} := 0$.
if ($p = 1$)
    $uplsp(t, 1, m, 0)$.

Table 5.9: Evaluating $uplsp(t, 1, \cdot, \cdot)$ where $1 \leq t < T$

a setup cost based rule

$$priority_{jt} \stackrel{def}{=} \frac{1}{s_j}, \tag{5.18}$$

or a capacity demand oriented rule

$$priority_{jt} \stackrel{def}{=} \sum_{\tau=1}^{t} p_j d_{j\tau}^{L4L}. \tag{5.19}$$

A dynamic rule can be given as

$$priority_{jt} \stackrel{def}{=} \frac{h_j CD_{jt}}{s_j} \tag{5.20}$$

where CD_{jt} is defined as in (5.13) using (5.15) to determine $\hat{t}$. Note, CD_{jt} and hence $priority_{jt}$ cannot be computed before the enumeration starts. But, since CD_{jt} is a cumulative value, it can efficiently be computed while moving stepwise from period to period adding up those demands which are not scheduled.

---

if ($j_{m0} \neq 0$)
    $mask_{j_{m0}1} := mask_{j_{m0}1} + p$.
if ($m = M$)
    $uplsp(0, 0, 1, 1)$.
else
    $uplsp(0, 1, m + 1, 1)$.
if ($j_{m0} \neq 0$)
    $mask_{j_{m0}1} := mask_{j_{m0}1} - p$.

Table 5.10: Evaluating $uplsp(0, 1, \cdot, \cdot)$

We use the dynamic rule (5.20) to compute priority values in our implementation. Its interpretation is that for not fulfilling future demand extra holding costs are charged. If setup costs are low, this tends to give bad solutions. But, if setup costs are high, building lots tends to be a good idea.

5.2.4 Bounding Rules

The enumeration scheme presented in Subsection 5.2.2 performs a complete enumeration and uses no insight information to prune the search tree. Hence, we should develop some bounding rules to reduce the computational effort.

First, let us consider the set of items to choose among in order to fix the setup state of a machine m. Both, in Table 5.7 and in Table 5.9, we defined $ITEMSET_{mt} = \mathcal{J}_m$ for initialization. Since switching the setup state is necessary only if production takes place, this item set can usually be chosen smaller. At the end of period T (Table 5.7) a machine m is set up for the item that is produced last on that machine. Any item for which external demand occurs could be that item. If no external demand occurs for an item, it can only be the last one on a machine if there is no successor item sharing the same machine. Hence,

$$ITEMSET_{mT} = \left\{j \in \mathcal{J}_m \mid \sum_{t=1}^{T} d_{jt} > 0\right\} \cup \left\{j \in \mathcal{J}_m \mid \{i \in \bar{\mathcal{S}}_j \mid m_i = m_j\} = \emptyset\right\} \quad (5.21)$$

is a valid choice. For periods t where $1 \leq t < T$ (Table 5.9) it is sufficient to consider those items only for which demand occurs in period t or in period $t+1$, or for which the machine is also set up in period $t+1$. The latter condition allows keeping the setup state up. More formally,

$$ITEMSET_{mt} = \{j \in \mathcal{J}_m \mid CD_{jt} > 0\} \cup \left\{j_{m(t+1)}\right\} \quad (5.22)$$

defines the initialization of $ITEMSET_{mt}$.

The production mask matrix is completely enumerated when following the lines of the *uplsp*–procedure. But, since a positive entry $mask_{jt}$ must only be considered if there is (future) demand for item j that is not been met, we can reduce the computational effort as follows: In Table 5.8 we add

if $(p = 1$ and $CD_{j_{mt}t} = 0)$ $p = 0$

at the very beginning to skip the consideration of the value $p = 1$. In Table 5.9 we simply add

if $(p = 1)$ $ITEMSET_{mt} := ITEMSET_{mt} \setminus \{j \in \mathcal{J}_m \mid CD_{j(t+1)} = 0\}$

right behind the initialization of $ITEMSET_{mt}$. As a consequence, the production masks being enumerated actually are binary[7] matrices now where positive entries in periods $1 < t \leq T$ do indeed represent that production takes place.

Another way to speed up the enumeration is to detect intermediate states which cannot lead to any feasible solution. For notational convenience, let

$$\mathcal{J}^{+}_{mt} \stackrel{def}{=} \{j \in \mathcal{J}_m \mid CD_{jt} > 0\} \quad (5.23)$$

[7]Only in period 1 production mask entries may have the value 2 which is due to the scheme in Table 5.10. Note, this does not mean unnecessary overhead in period 1, because the code in Table 5.10 does not enumerate the values for the parameter p.

denote the set of items which share machine m and for which there exists a positive cumulative demand in period t. On the basis of this,

$$\mathcal{J}_{mt}^{++} \stackrel{def}{=} \{j \in \mathcal{J}_m \cap \bar{\mathcal{P}}_i \mid i \in \bigcup_{\hat{m}=1}^{M} \mathcal{J}_{\hat{m}t}^{+}\} \cup \mathcal{J}_{mt}^{+} \tag{5.24}$$

defines a set of items which will have to be scheduled on machine m in periods 1 to t. This is true, because we have assumed no initial inventory. Owing to a unique setup state at the end of each period, at the beginning of the procedure given in Table 5.8 we therefore test

$$\mid \mathcal{J}_{mt}^{++} \mid > t+1 \tag{5.25}$$

which when true indicates that no feasible solution can be found any more. Similarly, when entering the code given in Tables 5.9 and 5.10 we check for

$$\mid \mathcal{J}_{m(t+1)}^{++} \mid > t+1 \tag{5.26}$$

and initiate a backtracking step depending on its outcome. Since we face multi–level gozinto–structures, it may happen that items causing internal demand for preceding items with respect to positive lead times do not fit into the remaining time window. More precisely, if we enter the code given in Table 5.8 and if there is an item

$$j \in \bigcup_{\hat{m}=1}^{M} \mathcal{J}_{\hat{m}t}^{+} \text{ where } dep_j \geq t \tag{5.27}$$

then the current node is fathomed. Analogously, when the code in Tables 5.9 and 5.10 is entered and if there is an item

$$j \in \bigcup_{\hat{m}=1}^{M} \mathcal{J}_{\hat{m}t}^{+} \text{ where } dep_j > t \tag{5.28}$$

then the current node is fathomed, too.

Eventually, we can prune the search tree on the basis of cost criteria. Let $costs^{PS}$ be the sum of setup and holding costs for a partial schedule ranging from periods $t+1$ to T. When initialized with zero when we start the enumeration, i.e. $costs^{PS} = 0$, it can easily be computed as we pass through the search tree. More precisely, consider what is done in Table 5.9. After j_{mt} is selected, we compute

$$costs^{PS} := costs^{PS} + s_{j_{m(t+1)}} \tag{5.29}$$

if $j_{mt} \neq j_{m(t+1)}$, because a setup then takes place in period $t+1$. Having added p to $mask_{j_{mt}(t+1)}$, we compute

$$costs^{PS} := costs^{PS} + \sum_{j \in \mathcal{J}_m} h_j CD_{j(t+1)} \tag{5.30}$$

for holding the cumulative demand one (more) period in inventory. Note, if $p = 1$ and thus $mask_{j_{mt}(t+1)} > 0$, then $CD_{j_{mt}(t+1)}$ evaluates to zero. Since backtracking may occur, we must not forget to perform the reverse operations

$$costs^{PS} := costs^{PS} - s_{j_{m(t+1)}} - \sum_{j \in \mathcal{J}_m} h_j CD_{j(t+1)}. \tag{5.31}$$

A valid position to execute this operation is, for example, right before the value p is subtracted from $mask_{j_{mt}(t+1)}$. The same operations can be performed within the code given in Table 5.10. The only difference is that there is no choice for j_{m0} which is the initial setup state. But, this does not affect the determination of $costs^{PS}$. Imagine that we have a lower bound for the minimum costs incurred in periods 1 to t, say, $lowerbound_{1t}$. Furthermore, suppose that we have an upper bound for the optimum solution of the overall instance, say, *upperbound*. Then, once we have increased $costs^{PS}$, we evaluate

$$lowerbound_{1t} + costs^{PS} > upperbound \tag{5.32}$$

which when true indicates that the choice of j_{mt} does not give an optimum solution. Hence, we can immediately loop back to select another setup state (or, if $t = 0$, initiate backtracking). What is left now is the discussion of determining *upperbound* and $lowerbound_{1t}$. The former follows standard ideas. Starting with an initial value (which might be infinity) we update its value whenever we reach a terminal node that improves the current best upper bound. A better initial value than infinity can be computed with any of the heuristics for the PLSP–MM which will be described in Chapter 6. For computing $lowerbound_{1t}$ we use a new though simple idea. We just solve the U–PLSP–instance that emerges from the original U–PLSP–instance when restricting our attention to the first t periods only. As a solution method we use the B&B–procedure again. Since this idea can be carried on recursively, we have to solve a sequence of U–PLSP–instances. The working principle should be made a bit more clear now: Set $lowerbound_{10} = 0$. First, we solve a U–PLSP–instance which is the instance that emerges from the original instance when we consider period 1 only. The result gives $lowerbound_{11}$. Then, we solve a U–PLSP–instance which consists of the first two periods of the original instance which yields $lowerbound_{12}$. This goes on until an instance with T periods is eventually solved. The instance with T periods is the original instance which gives the desired result. The remarkable point to note here is that when we are about to solve an instance with, say, t periods, we make use of $lowerbound_{11}, \ldots, lowerbound_{1(t-1)}$ which are previously computed. Since many instances have some periods with no external demand (see the discussion of macro and micro periods in Chapter 4), the stepwise procedure can be accelerated in most cases. If period t is a period with no external demand then solving the instance consisting of the first t periods gives the same objective function value as solving the instance consisting of the first $t-1$ periods. Hence, if $\sum_{j=1}^{J} d_{jt} = 0$ then choose $lowerbound_{1t} = lowerbound_{1(t-1)}$.

5.2.5 Experimental Evaluation

To study the performance of the presented B&B–procedure, we solve the small PLSP–instances which are defined in Section 4.2. Tables A.7 to A.9 in the Appendix A provide the results. For each parameter level combination we find the average deviation from the optimum result, the worst case result, and the number of instances for which a feasible solution exists. The deviation of the lower bound of an instance from its optimum solution is computed as defined in (5.12).

For a detailed analysis of the deviation data, we aggregate the results to see whether or not certain parameter levels have a significant impact on the performance. To start with, let us consider the number of machines first. Table 5.11 gives more insight. As we can see, increasing the number of machines reduces the average performance remarkably. The reason for this is not obvious. A possible explanation might be the following: On the one hand, producing a lot to fulfill a demand may last several periods if capacities are scarce, but takes one period if the capacity constraints are relaxed. Hence, the total holding costs tend to be underestimated. On the other hand, more machines tend to result in lower total holding costs, because items which do not share a common machine may be produced in parallel and need not be sequenced. Both together might explain the observed result, because the underestimation of holding costs is relatively high if total holding costs are low.

$M=1$	$M=2$
-9.70	-15.19

Table 5.11: The Impact of the Number of Machines on the Lower Bound

Next, we are interested in the impact of the gozinto–structure complexity. Table 5.12 shows the results. We cannot state that there is any effect.

$\mathcal{C}=0.2$	$\mathcal{C}=0.8$
-12.34	-12.57

Table 5.12: The Impact of the Gozinto–Structure Complexity on the Lower Bound

Table 5.13 reveals whether or not the demand pattern plays a role in performance changes. It can be seen that sparse demand matrices give poor results. Demand matrices with many non–zero entries give the best results with less than a 10% deviation on average. A reason for this phenomenon seems to be the fact that lot sizing is more important if there are many demands which can be grouped together in order to form a lot and thus to safe setup costs. These saving opportunities are detected by the B&B–method, too.

$(T_{macro}, T_{micro}, T_{idle}) =$		
$(10, 1, 5)$	$(5, 2, 2)$	$(1, 10, 0)$
-9.88	-11.20	-16.11

Table 5.13: The Impact of the Demand Pattern on the Lower Bound

If the cost structure is important for a good performance can be read in Table 5.14. Though differences are not dramatic, there is a tendency for instances with high setup costs to have a better lower bound than instances with low setup costs.

$COSTRATIO =$		
5	150	900
-13.42	-12.05	-11.89

Table 5.14: The Impact of the Cost Structure on the Lower Bound

Since we relax the capacity constraints, the capacity utilization is expected to have a significant impact on the performance of the B&B–method. As Table 5.15 indicates, this is indeed the case. For $U = 30$ we have an average deviation of -5.29% which is a fairly good result. But, for $U = 70$ the average deviation is -21.27% which is quite poor, although it is still better than the LP–relaxation. As a result, we have that the capacity utilization is the most significant factor for the computation of a lower bound via a capacity relaxation. Remarkable to note, the average deviation for $U = 30$ is not very close to zero which indicates that even a low capacity utilization does not make the multi–level lot sizing and scheduling problem be an easy–to–solve problem.

$U = 30$	$U = 50$	$U = 70$
-5.29	-11.73	-21.27

Table 5.15: The Impact of the Capacity Utilization on the Lower Bound

The run–time performance of the B&B–method varies very much. For the instances with $M = 1$, it takes between 0 to 53 CPU–seconds to solve them where zero time actually means that the run–time is too small to be measured. Most of the instances with $M = 2$ can also be solved within 60 CPU–seconds. Almost all instances can be solved with a time limit of 900 CPU–seconds. Three outliers need a bit less than 3,600 CPU–seconds.

In summary, the overall average deviation from the optimum results is a -12.46% deviation for our test–bed. This result decidedly outperforms the outcome of the LP–relaxation approaches. Hence, the proposed B&B–method provides a means to compute lower bounds for small instances within reasonable time. Lower bounds for medium– to large–size instances can, however, not be determined due to the exponential growth of the computational effort.

5.3 Lagrangean Relaxation

In the preceding section we removed the set of capacity constraints which complicate solving a PLSP–MM–instance. For the resulting problem we then developed a tailor–made procedure to find lower bounds. A bit more sophisticated than just removing complicating constraints is the general idea of replacing constraints with a penalty term in the objective function. The trick is to define a penalty term which — for minimization problems — increases the objective function value if the removed constraints are violated. Such an approach which additionally guarantees to give lower bounds for the original problem is known as Lagrangean relaxation. A nice introduction to it is given in [Fis85] (see also [Geo74, Fis81]).

Subsequently, we will describe how to apply a Lagrangean relaxation to the PLSP–MM in order to get a lower bound. Again, the capacity constraints are considered as the complicating constraints. We can therefore use the B&B–procedure described in the preceding section. In contrast what was done above, we now minimize the objective function

$$\sum_{j=1}^{J}\sum_{t=1}^{T}(s_j x_{jt} + h_j I_{jt}) - \sum_{m=1}^{M}\sum_{t=1}^{T}\lambda_{mt}^{(k)}\Delta capacity_{mt}^{(k)} \tag{5.33}$$

where

$$\Delta capacity_{mt}^{(k)} \stackrel{def}{=} C_{mt} - \sum_{j\in\mathcal{J}_m} p_j q_{jt}^{(k)} \qquad \begin{array}{l} m = 1,\ldots,M \\ t = 1,\ldots,T \end{array} \tag{5.34}$$

gives a negative value, if the capacity of a machine m in period t is exhaustively required. For the moment it suffices here to know that the values $\lambda_{mt}^{(k)}$ are non–negative parameters — so–called Lagrangean multipliers.

A procedure based on Lagrangean relaxation proceeds iteratively and the upper index (k) simply counts the iterations. Roughly speaking, starting with initial values for the Lagrangean multipliers, the resulting instance is solved. Afterwards, the multipliers are modified and a new iteration starts to solve the instance again. This goes on until a stopping criterion is met.

In our context, we begin with

$$\lambda_{mt}^{(1)} = 0 \qquad \begin{array}{l} m = 1,\ldots,M \\ t = 1,\ldots,T \end{array} \tag{5.35}$$

which actually means that we solve an U–PLSP–instance which is exactly the same as the one concerned in the preceding section.

After each iteration k we update the Lagrangean multipliers using

$$\lambda_{mt}^{(k+1)} = \max\left\{0, \lambda_{mt}^{(k)} - \delta^{(k)}\frac{(UB^* - LB^*)\Delta capacity_{mt}^{(k)}}{\sum_{\hat{m}=1}^{M}\sum_{\hat{t}=1}^{T}(\Delta capacity_{\hat{m}\hat{t}}^{(k)})^2}\right\} \tag{5.36}$$

for $m = 1, \ldots, M$ and $t = 1, \ldots, T$ as it is adviced in [HeWoCr74]. UB^* and LB^* are used to denote the smallest known upper bound and the largest known lower bound, respectively, for the PLSP–MM–instance under concern. Note, UB^* can be determined with one of the heuristics described in later chapters. LB^* can be chosen zero initially, and be updated after each iteration that gives a higher objective function value than LB^*. The parameter $\delta^{(k)}$ is a positive value where, again, [HeWoCr74] helps: Initially,

$$\delta^{(1)} = 2. \tag{5.37}$$

For $k > 1$, if LB^* has not increased within the last, say, $DELTAITER$ iterations then

$$\delta^{(k+1)} = \frac{\delta^{(k)}}{2}, \tag{5.38}$$

otherwise,

$$\delta^{(k+1)} = \delta^{(k)}. \tag{5.39}$$

The parameter $DELTAITER$ is chosen by the user.

The stopping rule to terminate the iteration process depends on several user specified parameters which are defined in Table 5.16 and which are tested after each iteration. If one of these criteria is fulfilled, the Lagrangean relaxation procedure stops giving LB^* as a lower bound. By construction, this lower bound is greater than or equal to the lower bound computed in the preceding section when we solved U–PLSP–instances.

It should be stressed that the B&B–method described in the preceding section can be used almost unchanged as a submodule. All that needs to be modified is the evaluation of production plans which now must use formula (5.33). The application of the cost–oriented bounding rule needs to compute

$$costs^{PS} := costs^{PS} + \sum_{j \in \mathcal{J}_m} (h_j CD_{j(t+1)} - \lambda_{mt}^{(k)} \Delta capacity_{mt}^{(k)}) \tag{5.40}$$

instead of (5.30). The reverse operations (5.31) must be adapted likewise. Remember Subsection 5.2.1 for determining the values $q_{jt}^{(k)} (= q_{jt})$ which in turn are needed for computing the values $\Delta capacity_{mt}^{(k)}$.

Symbol	Definition
$MAXGAP$	Terminate if $UB^* - LB^* \leq MAXGAP$.
$MAXITER$	Terminate if $MAXITER$ iterations are performed.
$MINLAMBDA$	Terminate if $\lambda_{mt}^{(k+1)} \leq MINLAMBDA$ for all $m = 1, \ldots, M$ and $t = 1, \ldots, T$.
$TIMELIMIT$	Terminate if the total run–time is greater than $TIMELIMIT$ seconds.

Table 5.16: Stopping Criteria for the Lagrangean Relaxation Approach

5.3.1 Experimental Evaluation

To study the performance of the Lagrangean relaxation procedure we solve the 1,033 small PLSP–MM–instances defined in Section 4.2. The method parameters are chosen as follows: $DELTAITER$ $= 5$, $MAXGAP = 0.001$, $MAXITER = 1,000$, $MINLAMBDA$ $= 0.0001$, and $TIMELIMIT = 3,600$. Tables A.10 to A.12 in the Appendix A provide the details of the outcome of the computational test. The information given for each parameter level combination has the same meaning as in earlier sections, i.e. the average deviation from the optimum results, the worst case result, and the number of instances that are solved. Again, the deviation of the lower bound for an instance from the optimum result is computed using (5.12).

The discussion of the results on the basis of aggregated data begins with the impact of the number of machines on the performance. Table 5.17 shows the numbers. As for the U–PLSP solution, the findings are that the average gap is positively correlated with the number of machines.

The effect of changing the complexity of the gozinto–structure is analyzed in Table 5.18. Although differences are not dramatic, gozinto–structures which are more complex tend to result in slightly larger deviations.

$M = 1$	$M = 2$
-3.82	-7.34

Table 5.17: The Impact of the Number of Machines on the Lower Bound

$\mathcal{C} = 0.2$	$\mathcal{C} = 0.8$
-5.26	-5.92

Table 5.18: The Impact of the Gozinto–Structure Complexity on the Lower Bound

The demand pattern is the focus of interest in Table 5.19. While there is a strong tendency for the U–PLSP to yield large deviations if the demand matrix is sparsely filled, the Lagrangean relaxation remedies this phenomenon and now gives best results for this case. However, there is not a clear tendency that more demand entries reduce the performance, because the results for the parameter level $(T_{macro}, T_{micro}, T_{idle}) = (10, 1, 5)$ are better than for $(T_{macro}, T_{micro}, T_{idle}) = (5, 2, 2)$.

$(T_{macro}, T_{micro}, T_{idle}) =$		
$(10, 1, 5)$	$(5, 2, 2)$	$(1, 10, 0)$
-5.59	-6.11	-5.11

Table 5.19: The Impact of the Demand Pattern on the Lower Bound

Table 5.20 reveals the impact of the cost structure on the performance. In contrast to the results in the preceding section, we now have that low setup costs give the best results on average. The average deviation is positively correlated with the ratio of setup and holding costs.

$COSTRATIO =$		
5	150	900
-4.58	-5.18	-7.00

Table 5.20: The Impact of the Cost Structure on the Lower Bound

For the capacity utilization we observe again that a high utilization results in a significantly worse deviation on average than a low utilization. Table 5.21 makes this apparent.

$U = 30$	$U = 50$	$U = 70$
-2.20	-4.81	-10.23

Table 5.21: The Impact of the Capacity Utilization on the Lower Bound

The average deviation for all 1,033 instances is -5.59% which is a satisfying result.

5.4 Summary of Evaluation

The computational study reveals that the we obtain no satisfying lower bounds by solving the LP–relaxation of the original PLSP–model formulation or a simple plant location formulation. Hence, a B&B–method is proposed to solve the uncapacitated PLSP optimally. The lower bound that we get this way is decidedly better. A Lagrangean relaxation of the capacity constraints improves these bounds. Table 5.22 summarizes the overall average results.

	Average Gap
LP–Relaxation of the PLSP–Model	-48.20
LP–Relaxation of the Simple Plant Location Model	-42.49
Solution of the Uncapacitated PLSP	-12.46
Lagrangean Relaxation of the Capacity Constraints	-5.59

Table 5.22: Summary of Lower Bounding Methods

Chapter 6

Multiple Machines

This chapter discusses several heuristic solution procedures for the PLSP–MM. We briefly report about some unsuccessful attempts in Section 6.1. A backward oriented construction scheme is underlying all methods described in here. Hence, in Section 6.2 we provide those construction principles that are common to all heuristics. Afterwards, we refine the presented concept to give a randomized regret based sampling procedure in Section 6.3, a cellular automaton in Section 6.4, a genetic algorithm in Section 6.5, a disjunctive arc based tabu search method in Section 6.6, and a so–called demand shuffle heuristic in Section 6.7. Section 6.8 provides a summary of the computational studies and compares the procedures. In Section 6.9 we run some tests with large instances to reveal the applicability to real–world instances.

6.1 Unsuccessful Attempts

While developing heuristics for the PLSP, not every idea that came to our mind turned out to be a good one, of course. Some of the approaches that failed or at least did not amaze in our studies shall be briefly reported in this section without giving too much detail or any computational results.

All unsuccessful attempts listed below were made when having a single–machine variant of the PLSP–MM under concern. The test–bed that was used consisted of small instances with five items and 10 periods only.

6.1.1 LP–Based Strategies

A set of fairly general approaches to attack problems for which MIP–model formulations exist are known as LP–based techniques. Loosely speaking, their idea is to solve a sequence of linear programs starting with the LP–relaxation. After each step, one or more (integer) variables are fixed by adding new equality constraints. Then, the resulting linear program is solved and so on until all integer variables are set [DiEsWoZh93, KuSaWaMa93, MaMClWa91].

In our context, the binary setup state variables were assigned a zero or a one step by step. Several strategies were tested. First, we tried so–called single–pass procedures where only one linear program (the LP–relaxation) was to be solved. Using some kind of rounding techniques assigned an integral value to all setup state variables. Second, we also tried so–called multi–pass procedures where we had an iterative process which repeatedly fixed some variables and solved linear programs. Different ways to choose the next set of variables to be fixed were tried out. For example, we moved on backwards fixing the setup states in period T first, then in period $T-1$, and so on until the process terminates after T iterations. Another idea was to choose the period for which the setup state is fixed next by means of priority rules. Both, the single–pass and the multi–pass procedures disappointed. Very often they did not even find a feasible solution, or, if a feasible solution was detected, gave a deviation from the optimum objective function value that was very poor. Using non–deterministic rules for fixing variables and repeating the process several times drastically increased the run-time effort, but gave no significant improvement.

As a consequence, we tested backtracking procedures moving on from period to period to find at least a feasible solution. Both, a complete enumeration and truncated versions were tested. The former approach gave the optimum result, but (as expected) its run–time performance was not encouraging. Only small instances could be solved within reasonable time. Truncated enumeration procedures which terminate upon a time limit suffered from the same shortcomings as single– and multi–pass procedures.

6.1.2 Improvement Schemes

A heuristic is called an improvement method if starting with a feasible solution the current solution is modified to give another feasible solution (not necessarily with a lower objective function value). This process may be repeated over and over again until some stopping criterion is met.

Some of the ideas we tried out could be termed as two–phase improvement schemes which, starting with a (feasible) setup state pattern (i.e. an assignment for all setup state variables), solved the remaining linear program either optimally with a standard solver or heuristically with a construction procedure. Afterwards, the setup state pattern was modified and the resulting linear program was solved again, and so on.

We were not able to find any other improvement method than the postprocessor defined in Subsection 3.3.8 that could be said to make a contribution. Maintaining feasibility turned out to be a very hard task.

As we will see, all heuristics described below are no improvement methods in the sense that we do not strictly move from one feasible solution to another feasible solution. Once that a feasible solution is found, it may happen that a next (intermediate) result is no longer feasible.

6.1.3 Data Perturbation

When we reviewed the literature in Chapter 2, it turned out that many authors use a modification of the cost parameters to take multi–level gozinto–structures into account when solving uncapacitated lot sizing problems by applying single–level heuristics level–by–level. A more general idea is discussed in [StWuPa93] where modifying the values of the problem parameters (data perturbation), solving the resulting instance, and evaluating the derived solution by making use of the original parameter values is proposed. While performing several iterations, a systematic search in the space of perturbed data values is advocated.

For the PLSP–MM we tried out a perturbation of setup and holding costs and used the heuristic to be described in Section 6.3. Neither a random choice of cost parameters nor a systematic local search using tabu search, simulated annealing, or genetic algorithms indicated that results can decidedly be improved.

6.1.4 Searching in the Method Parameter Space

Most heuristics depend on some parameters which guide the solution process. Similar to the search in the space of problem parameters (data perturbation), we can also imagine to perform a search in the space of the method parameters. Again, we used the heuristic to be described in Section 6.3 using tabu search, simulated annealing, and genetic algorithms for scanning the parameter space. Only small improvements were gained.

This result fits to earlier findings for a single–level PLSP heuristic where the parameter space was partitioned into small areas and a sequential test was used to decide which areas tend to give poor results [DrHa96].

6.2 Common Construction Principles

There is a generic construction scheme that forms the basis of all subsequent methods. It is a backward oriented procedure which schedules items period by period starting with period T and ending with period one. Similar to what was done in Chapter 5, we choose here a recurrent representation which enables us to develop the underlying ideas in a stepwise fashion. Now, let us assume that $construct(t, \Delta t, m)$ is the procedure to be defined. Analogously to the *uplsp*–procedure, $t + \Delta t$ is the period and m is the machine under concern. Again, $\Delta t \in \{0, 1\}$ where $\Delta t = 1$ indicates that the setup state for machine m at the beginning of period $t + 1$ is to be fixed next and $\Delta t = 0$ indicates that we already have chosen a setup state at the end of period t. Likewise to the *uplsp*–specification, j_{mt} will denote the setup state for machine m at the end of period t. Assume $j_{mt} = 0$ for $m = 1, \ldots, M$ and $t = 1, \ldots, T$ initially.

Before the construction mechanism starts, the decision variables y_{jt} and q_{jt} are assigned zero for $j = 1, \ldots, J$, $m = 1, \ldots, M$, and $t = 1, \ldots, T$. Remember, given the values for y_{jt} and q_{jt} the values for x_{jt} and I_{jt} are implicitly defined. Furthermore, assume auxiliary variables $\tilde{d}_{jt}$ and CD_{jt} for $j = 1, \ldots, J$ and $t = 1, \ldots, T$. The former ones represent the entries in the demand matrix and thus are initialized with $\tilde{d}_{jt} = d_{jt}$. The latter ones stand for the cumulative future demand for item j which is not been met yet. As we will see, the cumulative demand can be efficiently computed while moving on from period to period. For the sake of convenience we introduce $CD_{j(T+1)} = 0$ for $j = 1, \ldots, J$. The remaining capacity of machine m in period t is denoted as RC_{mt}. Initially, $RC_{mt} = C_{mt}$ for $m = 1, \ldots, M$ and $t = 1, \ldots, T$.

The initial call is *construct*$(T, 1, 1)$ and initiates the fixing of setup states at the end of period T. Table 6.1 gives all the details.

---

choose $j_{mT} \in \mathcal{I}_{mT}$.
if $(j_{mT} \neq 0)$
    $y_{j_{mT}T} := 1$.
if $(m = M)$
    construct$(T, 0, 1)$.
else
    construct$(T, 1, m + 1)$.

Table 6.1: Evaluating *construct*$(T, 1, \cdot)$

The choice of j_{mT} needs to be refined, but at this point we do not need any further insight and suppose that the selection is done somehow. All we need to know is that $\mathcal{I}_{mt} \subseteq \mathcal{J}_m \cup \{0\}$ for $m = 1, \ldots, M$ and $t = 1, \ldots, T$ is the set of items among which items are chosen. Item 0 is a dummy item which will be needed for some methods that will be discussed. We will return for a precise discussion in subsequent sections. As one can see, once a setup state is chosen for all machines at the end of period T, a call of *construct*$(T, 0, 1)$ is made. Table 6.2 provides a recipe of how to evaluate such calls.

---

for $j \in \mathcal{J}_m$
    $CD_{jt} := \min\left\{CD_{j(t+1)} + \tilde{d}_{jt}, \max\{0, nr_j - \sum_{\tau=t+1}^{T} q_{j\tau}\}\right\}$.
if $(j_{mt} \neq 0)$
    $q_{j_{mt}t} := \min\left\{CD_{j_{mt}t}, \frac{RC_{mt}}{p_{j_{mt}}}\right\}$.
    $CD_{j_{mt}t} := CD_{j_{mt}t} - q_{j_{mt}t}$.
    $RC_{mt} := RC_{mt} - p_{j_{mt}} q_{j_{mt}t}$.
    for $i \in \mathcal{P}_{j_{mt}}$
        if $(t - v_i > 0$ and $q_{j_{mt}t} > 0)$
            $\tilde{d}_{i(t-v_i)} := \tilde{d}_{i(t-v_i)} + a_{ij_{mt}} q_{j_{mt}t}$.
if $(m = M)$
    construct$(t - 1, 1, 1)$.
else
    construct$(t, 0, m + 1)$.

Table 6.2: Evaluating *construct*$(t, 0, \cdot)$ where $1 \leq t \leq T$

The situation when calling *construct*$(t, 0, m)$ is that the setup state j_{mt} has already been chosen. Remarkable to note, how easy it is to take initial inventory into account. This is due to the backward oriented scheme. Evaluating

$$\min\left\{CD_{j(t+1)} + \tilde{d}_{jt}, \max\{0, nr_j - \sum_{\tau=t+1}^{T} q_{j\tau}\}\right\} \tag{6.1}$$

makes sure that for an item j no more than the net requirement nr_j is produced. Note, cumulating the production quantities is an easy task which can be done very efficiently. Given the cumulative demand $CD_{j_{mt}t}$, production quantities $q_{j_{mt}t}$ can be determined with respect to capacity constraints. Afterwards, we simply update the $\tilde{d}_{jt}$–matrix to take internal demand into account and proceed. Table 6.3 describes how to evaluate *construct*$(t, 1, \cdot)$–calls.

choose $j_{mt} \in \mathcal{I}_{mt}$.
if $(j_{mt} \neq 0)$
    $y_{j_{mt}t} := 1$.
    if $(j_{mt} \neq j_{m(t+1)})$
        $q_{j_{mt}(t+1)} := \min\left\{CD_{j_{mt}(t+1)}, \frac{RC_{m(t+1)}}{p_{j_{mt}}}\right\}$.
        $CD_{j_{mt}(t+1)} := CD_{j_{mt}(t+1)} - q_{j_{mt}(t+1)}$.
        $RC_{m(t+1)} := RC_{m(t+1)} - p_{j_{mt}} q_{j_{mt}(t+1)}$.
        for $i \in \mathcal{P}_{j_{mt}}$
            if $(t + 1 - v_i > 0$ and $q_{j_{mt}(t+1)} > 0)$
                $\tilde{d}_{i(t+1-v_i)} := \tilde{d}_{i(t+1-v_i)} + a_{ij_{mt}} q_{j_{mt}(t+1)}$.
if $(m = M)$
    $construct(t, 0, 1)$.
else
    $construct(t, 1, m + 1)$.

Table 6.3: Evaluating $construct(t, 1, \cdot)$ where $1 \leq t < T$

These lines closely relate to what is defined in Table 6.2. Differences lie in the fact that a setup state is chosen for the end of period t but items are scheduled in period $t + 1$. For computing production quantities we must therefore take into account that item $j_{m(t+1)}$ may already be scheduled in period $t + 1$.

Note, the combination of what is given in Tables 6.2 and 6.3 enforces that every item j_{mt} that is produced at the beginning of a period $t + 1$ is also produced at the end of period t if there is any positive cumulative demand left. In preliminary tests not reported here we also found out that if capacity is exhausted, i.e. if $RC_{m(t+1)} = 0$ and $CD_{j_{m(t+1)}(t+1)} > 0$, it is best to choose $j_{mt} = j_{m(t+1)}$ in Table 6.3. In other words, lots are not split.[1] The reason why this turned out to be advantageous is that the setup state tends to flicker otherwise and thus the total sum of setup

[1]It is worth to be stressed that lot splitting could be easily integrated by *not* checking for exhausted capacity. All methods based on the described construction scheme may thus be adapted for lot splitting with minor modifications only.

costs tends to be high. In the rest of this chapter we assume that lots are not split.

Turning back to the specification of the *construct*–procedure, it remains to explain what shall happen when the first period is reached. Table 6.4 describes how to schedule those items in period 1 for which the machines are initially set up for. In contrast to what is given in Table 6.3 the initial setup state is known and thus needs not to be chosen.

---

if $(j_{m0} \neq j_{m1})$

$$q_{j_{m0}1} := \min\left\{CD_{j_{m0}1}, \frac{RC_{m1}}{p_{j_{m0}}}\right\}.$$

$$CD_{j_{m0}1} := CD_{j_{m0}1} - q_{j_{m0}1}.$$

if $(m = M)$

$\quad$ $construct(0, 0, 1)$.

else

$\quad$ $construct(0, 1, m + 1)$.

Table 6.4: Evaluating $construct(0, 1, \cdot)$

A call to $construct(0, 0, \cdot)$ terminates the construction phase. What is left is a final feasibility test where

$$nr_j = \sum_{t=1}^{T} q_{jt} \tag{6.2}$$

must hold for $j = 1, \ldots, J$ for being a feasible solution. Eventually, a feasible solution can be postprocessed as described in Subsection 3.3.8 and its objective function value can be determined.

If we have no initial inventory, the feasibility tests described in Subsection 5.2.4 are applied here as well to halt the execution before period 1 is reached. In addition to those tests, we can also perform a capacity check testing

$$\sum_{j \in \mathcal{J}_m} \sum_{i \in \{\bar{\mathcal{P}}_j \cup \{j\}\} \cap \mathcal{J}_m} p_i id_{ji} CD_{j(t+\Delta t)} > \sum_{\tau=1}^{t+\Delta t} C_{m\tau} \tag{6.3}$$

which must be false for $m = 1, \ldots, M$ if period $t + \Delta t$ is under concern and thus, when true, indicates an infeasible solution (if there is no initial inventory).

It should be emphasized again, that the construction scheme described above does not necessarily generate an optimum solution. It does not even guarantee to find a feasible solution if there exists one.

6.3 Randomized Regret Based Sampling

Now, having introduced the backward oriented construction scheme *construct*, a very straightforward idea is to run the construction phase over and over again while memorizing the current best plan until some stopping criterion (e.g. a certain number of iterations) is met. Note, this only makes sense if the *construct*–procedure works non–deterministically.

Here, the choice of setup states j_{mt} comes in again. If a stochastic selection rule is used then it is probable to have different results after each run of the construction phase.

Preliminary studies of ideas reported in this section are provided in [DrHa95, Haa94] for single–level, single–machine PLSP–instances and in [Kim96, Kim94a] for multi–level, single–machine PLSP–instances.

6.3.1 An Introduction to Random Sampling

The process of random sampling as done here is a Monte Carlo experiment where item numbers j_{mt} are repeatedly drawn at random out of a population $\mathcal{I}_{mt} \subseteq \mathcal{J}_m$ for $m = 1, \ldots, M$ and $t = 1, \ldots, T$. An underlying distribution function φ_{mt} is defined on the basis of a priority value $\pi_{jt} > 0$ that is assigned to each item j in the item set $\mathcal{I}_{mt}$.

Before we give any details of the definitions of φ_{mt} and π_{jt}, let us have a look at a taxonomy of sampling. Depending on the priority values, three important cases are worth to be highlighted. First, there is the general case with a probability to choose $j \in \mathcal{I}_{mt}$

being defined as

$$\varphi_{mt}(j) \stackrel{def}{=} \frac{\pi_{jt}}{\sum_{i \in \mathcal{I}_{mt}} \pi_{it}} \tag{6.4}$$

for $m = 1, \ldots, M$ and $t = 1, \ldots, T$ where

$$\sum_{j \in \mathcal{I}_{mt}} \varphi_{mt}(j) = 1 \tag{6.5}$$

holds. Since the priority values for the items may differ, this is called biased random sampling. By definition, if $j, i \in \mathcal{I}_{mt}$ and $\pi_{jt} > \pi_{it}$, then $\varphi_{mt}(j) > \varphi_{mt}(i)$.

Many authors apply biased random sampling procedures to many different kinds of application areas. A comprehensive overview of research activities is out of the scope of this book. See for instance [FeReSm94] where a so-called greedy randomized adaptive search procedure (GRASP) is introduced and used for finding maximum independent sets in graphs. Further references to GRASP-applications in the area of corporate acquisition of flexible manufacturing equipment, computer aided process planning, airline flight scheduling and maintenance base planning, and several other problems are given.

As a special case one might choose

$$\pi_{jt} = constant \tag{6.6}$$

for $j \in \mathcal{I}_{mt}$. This is called pure random sampling,[2] because all items $j \in \mathcal{I}_{mt}$ now have the same probability

$$\varphi_{mt}(j) = \frac{1}{\mid \mathcal{I}_{mt} \mid} \tag{6.7}$$

for $m = 1, \ldots, M$ and $t = 1, \ldots, T$.

Last, since items with a large priority values are preferred, one might compute a modified priority value

$$\varrho_{jt} \stackrel{def}{=} (\pi_{jt} - \min_{i \in \mathcal{I}_{mt}} \pi_{it} + \epsilon)^{\delta} \tag{6.8}$$

[2] Note that items $j \in \mathcal{J}_m \backslash \mathcal{I}_{mt}$ have a priority value $\pi_{jt} = 0$. So, we face a pure random sampling only if $\mathcal{I}_{mt}$ is known before the procedure starts, but not if $\mathcal{I}_{mt}$ depends on the history of the execution.

for each $j \in \mathcal{I}_{mt}$ where $\epsilon > 0$ and $\delta \geq 0$, and then define

$$\varphi_{mt}(j) \stackrel{def}{=} \frac{\varrho_{jt}}{\sum_{i \in \mathcal{I}_{mt}} \varrho_{it}} \tag{6.9}$$

for $m = 1, \ldots, M$ and $t = 1, \ldots, T$. For using such a kind of distribution function (in the context of project scheduling), Drexl coined the name regret based (biased) random sampling [Dre91, DrGr93]. The motivation for this name stems from the idea that the modified priority value ϱ_{jt} represents a measure for the regret not to choose the item j in period t. Hence, more emphasis is given on the differences of priority values π_{jt}. Both, ϵ and δ are method parameters that guide the sampling process.[3] For ϵ a small positive value should be chosen to make sure that every item $j \in \mathcal{I}_{mt}$ is assigned a positive modified priority value $\varrho_{jt} > 0$. The parameter δ amplifies (smoothes) differences in the priority values if $\delta > 1$ $(0 \leq \delta < 1)$.

6.3.2 Randomized Regret Based Priority Rules

The heart of a random sampling procedure are the priority values π_{jt} that are used to define the distribution function φ_{mt} for $m = 1, \ldots, M$ and $t = 1, \ldots, T$. But before we can introduce priority values, we first need to define the set of items $\mathcal{I}_{mt}$ among which an item j_{mt} is to be chosen.[4] Promising candidates to set a machine m up for at the end of a period t are those items for which there is demand in either period t or, if $t < T$, period $t+1$. In addition to that, items with demand in periods earlier than period t might be promising as well. Choosing such items causes idle periods (during which the setup state can be kept up). Apparently, the cumulative production quantities of an item need not exceed its

[3]Using different parameters ϵ_{mt} and δ_{mt} for each machine m and each period t turned out to increase the run–time, but did not improve the results decidedly.

[4]Remember, in some cases which are described in Section 6.2 no selection is to be made, because we do not allow lot splitting.

net requirement. More formally,

$$\mathcal{I}_{mt} \stackrel{def}{=} (\{j \in \mathcal{J}_m \mid CD_{j(t+1)} + \tilde{d}_{jt} > 0\} \cup \{j \in \mathcal{J}_m \mid \sum_{\tau=1}^{t-1} \tilde{d}_{j\tau} > 0\})$$

$$\cap \{j \in \mathcal{J}_m \mid nr_j - \sum_{\tau=t+1}^{T} q_{j\tau} > 0\} \tag{6.10}$$

for $m = 1, \ldots, M$ and $t = 1, \ldots, T$. If $\mathcal{I}_{mt} = \emptyset$, we simply choose $j_{mt} = j_{m(t+1)}$ in periods $t < T$, or, if $t = T$, fix $j_{mT} \in \mathcal{J}_m$ by arbitration, e.g. the item with the lowest item index (which when turned out to be wrong is neatly be corrected by the postprocessor). Let us therefore assume $\mathcal{I}_{mt} \neq \emptyset$.

Using our problem understanding, two main aspects help us to find priority values. On the one hand, we like to have low–cost production plans. A priority value should therefore reflect cost criteria. On the other hand, it is quite hard to generate even a feasible solution. Thus, priority values should also consider sources of infeasibility. In combination, priority values should lead to cheap and feasible production plans.

We start with introducing two expressions that represent cost criteria. First, imagine that a machine m is not set up for an item j at the end of period t. This means that $CD_{j(t+1)}$ items must be stored in inventory for at least one additional period and thus causing additional holding costs. The expression

$$\pi^{I}_{jt} \stackrel{def}{=} \frac{h_j CD_{j(t+1)}}{\max\{s_i \mid i \in \mathcal{J}_m\}} \tag{6.11}$$

for $m = 1, \ldots, M$, $t = 1, \ldots, T$, and $j \in \mathcal{I}_{mt}$ is a measure for these costs.

Second, changing the setup state causes setup costs, or, stating this the other way around, not to select a certain item may save setup costs. Thus,

$$\pi^{II}_{jt} \stackrel{def}{=} \frac{s_j}{\max\{s_i \mid i \in \mathcal{J}_m\}} \tag{6.12}$$

for $m = 1, \ldots, M$, $t = 1, \ldots, T$, and $j \in \mathcal{I}_{mt}$ is a measure for the cost savings, if an item is not selected.

In addition to that, we now give two expressions that tend to avoid infeasibility. On the one hand, we take into account that items usually have preceding items which are to be manufactured in advance. The more preceding items there are, the more risky it is to shift production into early periods. The closer we move towards period 1 the more important is this aspect. The expression

$$\pi_{jt}^{III} \stackrel{def}{=} \frac{dep_j}{t+1-dep_j} \tag{6.13}$$

for $m = 1, \ldots, M$, $t = 1, \ldots, T$, and $j \in \mathcal{I}_{mt}$ thus makes it more probable to choose items with a large depth, especially in early periods. To be well–defined choose a denominator equal to one if $t+1 = dep_j$.

On the other hand, we consider the capacity usage. Producing an item j and all its preceding items requires a well–defined amount of capacity per machine. The bottleneck machine can then be defined as the machine with the highest ratio of capacity demand per available capacity. Items which cause a high capacity usage on the bottleneck machine should therefore have a high preference to be chosen. More formally,

$$\begin{aligned} \pi_{jt}^{IV} \stackrel{def}{=} & \left(CD_{j(t+1)} + \tilde{d}_{jt}\right) \\ & \cdot \max\left\{ \frac{\sum_{i \in (\mathcal{P}_j \cup \{j\}) \cap \mathcal{J}_m} p_i d_{ji}}{\sum_{\tau=1}^{t} C_{m\tau}} \mid m \in \{1, \ldots, M\} \right\} \end{aligned} \tag{6.14}$$

for $m = 1, \ldots, M, t = 1, \ldots, T$, and $j \in \mathcal{I}_{mt}$ estimates the capacity usage of the bottleneck machine.

As preliminary tests have revealed, using a combination of these four criteria seems to be a good strategy. For $m \in \{1, \ldots, M\}$, $t \in \{1, \ldots, T\}$, and $j \in \mathcal{I}_{mt}$ four cases may now occur where $j_{m(T+1)} = 0$ is assumed for notational convenience:

Case 1: $CD_{j(t+1)} + \tilde{d}_{jt} > 0$ and $j \neq j_{m(t+1)}$.

Case 2: $CD_{j(t+1)} + \tilde{d}_{jt} > 0$ and $j = j_{m(t+1)}$.

Case 3: $\sum_{\tau=1}^{t-1} \tilde{d}_{j\tau} > 0$ and $j \neq j_{m(t+1)}$ and Case 1 does not hold.

Case 4: $\sum_{\tau=1}^{t-1} \tilde{d}_{j\tau} > 0$ and $j = j_{m(t+1)}$ and Case 2 does not hold.

Depending on the case that holds, π_{jt} can now be defined as

$$\pi_{jt} \stackrel{def}{=} \begin{cases} \gamma_1 \pi_{jt}^{I} - \gamma_2 \pi_{jt}^{II} + \gamma_3 \pi_{jt}^{III} + \gamma_4 \pi_{jt}^{IV} & \textit{Case 1} \\ \gamma_1 \pi_{jt}^{I} + \gamma_3 \pi_{jt}^{III} + \gamma_4 \pi_{jt}^{IV} & \textit{Case 2} \\ \gamma_3 \pi_{jt}^{III} + \gamma_4 \pi_{jt}^{IV} & \textit{Case 3} \\ \gamma_2 \pi_{jt}^{II} + \gamma_3 \pi_{jt}^{III} + \gamma_4 \pi_{jt}^{IV} & \textit{Case 4} \end{cases} \quad (6.15)$$

where $\gamma_1, \ldots, \gamma_4$ are real–valued method parameters.[5] Cases 1 and 2 are defined as motivated above. Note, that in Case 2 the setup cost savings are of no interest, because the machine already is in the proper setup state. In the Cases 3 and 4, the holding cost criterion evaluates to zero. In contrast to the first two cases, the setup cost criterion now increases the priority value if the machine is in the correct setup state (Case 4) and is neglected in Case 3. This differs from what is done in Cases 1 and 2, although its interpretation, i.e. giving items for which the machine is already set up for a better chance to be selected, is the same. The reason for doing so is motivated by a desire to decide for idle periods in order to maintain the setup state (Case 4). This idea is supported by making the priority value for the item for which Case 4 holds large and lower the priority values for those items which apply to Case 1.

Using (6.9) in combination with (6.8) we now have a full specification for the selection rule we employ. If the method parameters are chosen at random, the solution procedure is called randomized regret based (biased random) sampling.

6.3.3 Tuning the Method Parameters

In our implementation we choose all method parameters, namely $\gamma_1, \ldots, \gamma_4$, ϵ, and δ, at random in each iteration of the sampling process. Rather than doing this at pure random over and over again without any history sensitivity, one should make use of the information gained during previous iterations. This is to say that

[5]Using different parameters $\gamma_{1mt}, \ldots, \gamma_{4mt}$ for each machine m and each period t turned out to increase the computational overhead without giving decidedly better results.

learning effects should steer the choice of the method parameters for values that tend to give good results. As reported in Section 6.1, for the randomized regret based procedure no amazing results were achieved. However, we like to discuss here one way to tune the method parameters.

Since all that will be said is valid for all parameters, let

$$parameter \in \{\gamma_1, \ldots, \gamma_4, \epsilon, \delta\} \tag{6.16}$$

denote any of the method parameters. Furthermore, suppose $n = 1, 2, \ldots$ is the number of the current iteration and $parameter^*$ is the value of the parameter when the current best (feasible) solution was found within the last $n-1$ iterations. Initially, $parameter^* = 0$ without loss of generality. In each iteration we randomly choose a parameter value, say pv, out of the interval of valid parameter values, say $[parameter_{min}, parameter_{max}]$, as specified by the user. If we would assign $parameter = pv$, we have a pure random choice without any learning.

A fundamental idea for tuning the method parameters is to intensify the search in those areas of the parameter space which are in the neighborhood of $parameter^*$. In our implementation, this is done using

$$parameter = parameter^* + \nu_n(pv - parameter^*) \tag{6.17}$$

where $\nu_n \in [0, 1]$ defines the neighborhood around the current best parameter value from which the parameter used in iteration n is to be selected. If $\nu_n = 1$ then we have a pure random choice of parameters again. And if $\nu_n = 0$ then we would have no choice at all, because in every iteration $parameter^*$ would be chosen again. Initially, we start with $\nu_1 = 1$. Note that in each iteration the parameter value lies indeed in the valid parameter space, i.e.

$$parameter_{min} \leq parameter \leq parameter_{max}. \tag{6.18}$$

Without any significant effort, we can count the number of iterations done so far that resulted in improved solutions. Let this number be denoted as $improvements$. Starting with a counter $improvements = 0$, its value is set to one if the first feasible solution is found, and its value is increased by one every time a new current best solution is found. Having this information, we compute

$$\nu_{n+1} = \nu_n \tag{6.19}$$

if iteration n did not improve the current best solution and

$$\nu_{n+1} = \frac{1}{improvements} \tag{6.20}$$

if iteration n resulted in an improvement. Note, because the counter *improvements* monotonically increases from iteration to iteration, the value of ν_n declines during run–time and intensifies the search in a more and more narrow subspace of the parameter space.

In preliminary experiments we found out, that the intensification procedure described so far quite often traps into local optima, because the value ν_n declines too fast. The final outcome is often worse than what comes out with a pure random choice of parameter values in every iteration. Thus, we decided to choose $\nu_{n+1} = \nu_n$ for $n \leq NOINTENSIFY$ where $NOINTENSIFY$ is specified by the user.[6] Nevertheless, the counter *improvements* is maintained during the first $NOINTENSIFY$ iterations as well.

In addition to that, we found out that for some instances it is quite hard to find a feasible solution. In such cases, intensification turned out to be good. For instances for which feasible solutions are easily found intensification is less exciting. A counter for the number of iterations which revealed no feasible solution, namely $infeasible$, thus allows to decrease the parameter ν_n only if

$$\frac{infeasible}{n} > CRITICAL \tag{6.21}$$

where $CRITICAL \in [0,1]$ is a user–specified parameter.

[6] For instance, during tests with $J = 5$ and $T = 10$ we observed, that after 500 iterations we very often have results that are close to the results after 1,000 iterations when a pure random choice of parameter values is done [Kim94a]. So, we would suggest $NOINTENSIFY = 500$ for these instances.

6.3.4 Modifications of the Construction Scheme

The construction scheme given in Section 6.2 would work fine without any modification. However, suppose that we have chosen to set machine m up for item $j_{mt} \in \mathcal{I}_{mt}$ at the end of period t. Furthermore, assume that

$$j_{mt} \notin \{j \in \mathcal{J}_m \mid CD_{j(t+1)} + \tilde{d}_{jt} > 0\} \tag{6.22}$$

which implies that

$$j_{mt} \in \{j \in \mathcal{J}_m \mid \sum_{\tau=1}^{t-1} \tilde{d}_{j\tau} > 0\}. \tag{6.23}$$

Let $t_{j_{mt}}$ be the next period in which demand for item j_{mt} occurs, i.e.

$$t_{j_{mt}} = \max\{\tau \mid 1 \leq \tau < t \wedge \tilde{d}_{j_{mt}\tau} > 0\}. \tag{6.24}$$

Since the construction procedure strictly moves on from period to period choosing items again and again, it is quite unlikely to begin in period $t > t_{j_{mt}}$ and reach period $t_{j_{mt}}$ having no other item than j_{mt} be chosen for machine m at the end of periods $t_{j_{mt}}, t_{j_{mt}} + 1, \ldots, t$.

For this reason, we introduce the following modification: If an item j_{mt} fulfilling (6.22) is chosen then we update the $\tilde{d}_{jt}$–entries as described in Table 6.5.

---

for $j \in \mathcal{J}_m$
    $\tilde{d}_{jt_{j_{mt}}} := CD_{j(t+1)} + \sum_{\tau=t_{j_{mt}}}^{t} \tilde{d}_{j\tau}$.
    $CD_{j(t+1)} := 0$.
    for $\tau \in \{t_{j_{mt}} + 1, \ldots, t\}$
        $\tilde{d}_{j\tau} := 0$.

Table 6.5: Generating Idle Periods $t_{j_{mt}} + 1, \ldots, t$

This makes it unlikely to choose any other item than j_{mt} to be produced on machine m in the periods $t_{j_{mt}} + 1, \ldots, t-1$, because all demand that would require machine m is shifted to period $t_{j_{mt}}$.

Note, there is no guarantee to choose j_{mt} again and again. This makes sense, because after j_{mt} is selected at the end of period t it may happen that scheduling items on machines other than machine m causes updates of the $\tilde{d}_{j\tau}$–entries where $t_{j_{mt}} < \tau < t$. This demand has previously not been taken into account and hence we decide again if the setup state of machine m should indeed be kept up for item j_{mt}.

6.3.5 An Example

Consider the gozinto–structure given in Figure 6.1 and the parameters in Table 6.6. Furthermore, assume $M = 2$, $m_1 = m_4 = 1$, $m_2 = m_3 = 2$, and $C_{1t} = C_{2t} = 15$ for $t = 1, \ldots, 10$. For illustrating the construction of a production plan we do not need any information about setup and holding costs.

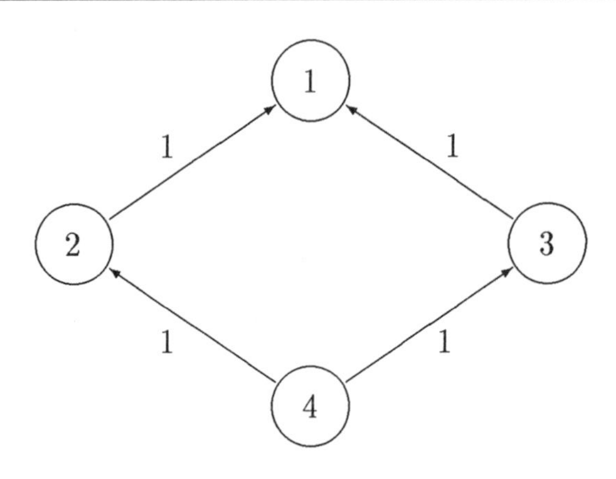

Figure 6.1: A Gozinto–Structure with Four Items

Table 6.7 provides a (part of a) protocol of one iteration of the sampling procedure where each row represents exactly one call of the *construct*–method. Column two shows the parameters that are passed to the *construct*–procedure. The entries $\vec{\tilde{d}}_{jt}$ are a tuple representation of the $\tilde{d}_{jt}$–values where for each item j the value $\tilde{d}_{jt}$

d_{jt}	$t=1$	...	5	6	7	8	9	10	p_j	v_j	y_{j0}	I_{j0}
$j=1$					20			20	1	1	1	0
$j=2$									1	1	1	0
$j=3$									1	1	0	0
$j=4$						5			1	1	0	0

Table 6.6: Parameters of the Example

can be found at position j in the tuple. Analogously, $\vec{CD}_{j(t+\Delta t)}$ are the cumulative demand values of interest. The column $\mathcal{I}_{mt}$ provides the set of items from which we can choose a setup state. The column j_{mt} gives the outcome of a random choice. For the sake of simplicity, we leave out the priority values which would not deepen the understanding. The production quantities as determined by the construction procedure are given, too. Figure 6.2 shows a possible result when starting with the lines given in the protocol.

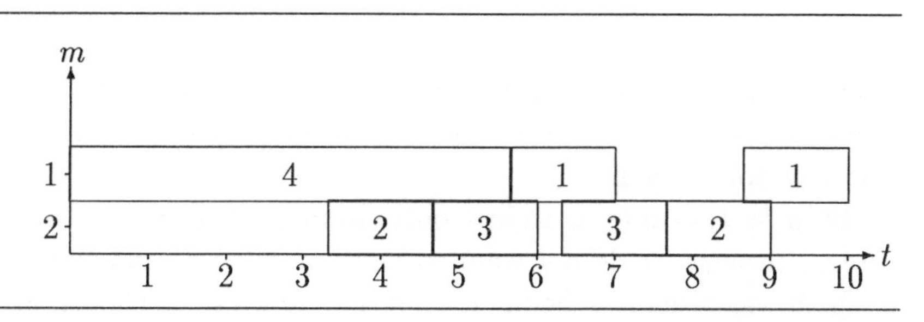

Figure 6.2: A Possible Outcome of the Run in the Protocol

Some points of interest are worth to be highlighted:

Step 1: $\mathcal{I}_{1T}$ contains both, item 1 and item 4, because we have demand for item 1 in period 10 and demand for item 4 in period 8.

Step 2: The set $\mathcal{I}_{2T}$ is empty, because we neither have any demand for item 2 nor for item 3. By arbitration, we choose item 2. Note, if we would choose item 3, but keep the rest of the protocol

Step	$(t, \Delta t, m)$	$\vec{\tilde{d}}_{jt}$	$\vec{CD}_{j(t+\Delta t)}$	$\mathcal{I}_{mt}$	j_{mt}	$q_{j_{mt}(t+\Delta t)}$
1	(10,1,1)	(20,0,0,0)	(0,0,0,0)	$\{1,4\}$	1	
2	(10,1,2)	(20,0,0,0)	(0,0,0,0)	$\emptyset$	2	
3	(10,0,1)	(20,0,0,0)	(20,0,0,0)			$q_{1T} = 15$
4	(10,0,2)	(20,0,0,0)	(5,0,0,0)			$q_{2T} = 0$
5	(9,1,1)	(0,15,15,0)	(5,0,0,0)	$\{1\}$	1	
6	(9,1,2)	(0,15,15,0)	(5,0,0,0)	$\{2,3\}$	2	
7	(9,0,1)	(0,15,15,0)	(5,0,0,0)			$q_{19} = 5$
8	(9,0,2)	(0,15,15,0)	(0,15,15,0)			$q_{29} = 15$
9	(8,1,1)	(0,5,5,20)	(0,0,15,0)	$\{1,4\}$	1	
10	(8,1,2)	(0,5,5,0)	(0,0,15,0)	$\{2,3\}$	2	
11	(8,0,1)	(0,5,5,0)	(0,0,15,0)			$q_{18} = 0$
12	(8,0,2)	(0,5,5,0)	(0,5,20,0)			$q_{28} = 5$
13	(7,1,1)	(20,0,0,25)	(0,0,20,0)	$\{1,4\}$	1	
14	(7,1,2)	(20,0,0,25)	(0,0,20,0)	$\{3\}$	3	$q_{38} = 10$
...						

Table 6.7: A Protocol of the Construction Scheme of the Regret Based Procedure

unchanged choosing item 2 in Step 6, then it seems to have an unnecessary setup in period 10. Due to the postprocessing stage, this will not happen.

Step 5: $\mathcal{I}_{19}$ contains item 1 only. Item 4 would not be a valid choice, because we do not want to have lot splitting and thus item 1 which was already scheduled in period 10 must be chosen again in period 9.

Step 8: Updating the demand matrix yields $\tilde{d}_{48} = 20$ (see Step 9) which is computed by adding up 5 (the external demand) and 15 (the internal demand).

Step 9: Although there is no demand for item 1 in period 8 ($\tilde{d}_{18} = 0$) and in period 9 ($CD_{19} = 0$), we choose item 1 again for which we have a demand in period 7. This should introduce idle periods and thus we shift the demand for item 4 from period 8 to period 7.

Step 10: Although the lot of item 2 which is scheduled in period

9 needs all the available capacity, all cumulative demand for item 2 fits into period 9 and does not require any capacity in period 8. Hence, $\mathcal{I}_{28}$ contains items 2 and 3, but not item 2 only.

Step 14: The machine 2 is set up for item 3 at the end of period 7. Since there is a positive cumulative demand for item 3 ($CD_{38} = 20$) and there is some free capacity left over in period 8 ($C_{28} - p_2 q_{28} = 15 - 5 = 10$), we can schedule 10 units of item 3 in period 8 which reduces the cumulative demand.

6.3.6 Experimental Evaluation

To test the performance of the randomized regret based sampling method we use the test–bed defined in Section 4.2. In each iteration the method parameters are drawn at random so that $\gamma_1, \ldots, \gamma_4 \in [0, 1]$, $\epsilon \in [0.0001, 0.1]$, and $\delta \in [0, 10]$. For tuning the method parameters we use $NOINTENSIFY = 500$ and $CRITICAL = 0.6$. Tables B.1 to B.3 in the Appendix B provide the results for 1,000 iterations of the sampling method. For each parameter level combination we find the average deviation of the upper bound from the optimum solution where for each instance the deviation is computed by

$$deviation \stackrel{def}{=} 100 \frac{UB - OPT}{OPT} \tag{6.25}$$

with UB being the upper bound and OPT the optimum objective function value. Moreover, the tables show the worst case deviation and the number of instances for which a feasible solution is found, too.

An analysis shall now be done on the basis of aggregated data to see whether or not certain parameter levels have an impact on the performance of the solution procedure. For each parameter level we give three performance measures. The average deviation from the optimum results is, perhaps, the most important. But, the infeasibility ratio which is the percentage of instances for which no feasible solution can be found although such one exists is a good indicator of what makes instances hard to solve, too. The average run–time performance is also shown where all values are given in CPU–seconds.

The effect of changing the number of machines is studied in Table 6.8. As we see, both, the average deviation from optimum and the average run–time significantly increase with the number of machines. However, more machines make finding a feasible solution easier.

	$M = 1$	$M = 2$
Average Deviation	8.90	11.69
Infeasibility Ratio	11.46	7.92
Average Run–Time	0.39	0.51

Table 6.8: The Impact of the Number of Machines on the Performance

Varying the gozinto–structure complexity also effects the performance of the regret based solution procedure. Table 6.9 provides the details. A high complexity results in larger deviations from the optimum than a low complexity. It also raises the infeasibility ratio decidedly. A minor impact is on the run–time performance where gozinto–stuctures with a greater complexity have a small tendency to increase the computational effort very slightly on average.

	$\mathcal{C} = 0.2$	$\mathcal{C} = 0.8$
Average Deviation	9.23	11.53
Infeasibility Ratio	7.05	12.40
Average Run–Time	0.44	0.47

Table 6.9: The Impact of the Gozinto–Structure Complexity on the Performance

Table 6.10 shows if the demand pattern affects the performance of the heuristic, too. We learn that a pattern with many non–zeroes in the demand matrix gives poor results. The average deviation from the optimum is positively correlated with the number of demands to be fulfilled. For sparsely–filled demand matrices the infeasibility ratio is below 1%, but increases dramatically for demand matrices with many entries. Differences in the run–time performance are not worth to be mentioned.

	$(T_{macro}, T_{micro}, T_{idle}) =$		
	$(10, 1, 5)$	$(5, 2, 2)$	$(1, 10, 0)$
Average Deviation	16.25	12.63	3.52
Infeasibility Ratio	14.53	14.20	0.84
Average Run–Time	0.48	0.44	0.44

Table 6.10: The Impact of the Demand Pattern on the Performance

The impact of the cost structure on the performance is analyzed in Table 6.11. It can be seen that for high setup costs the deviation of the upper bound from the optimum objective function value is on average significantly smaller than for low setup costs. The infeasibility ratios show that instances with very high or very low setup costs are not as easy to solve as instances with a balanced cost structure. The influence of different costs on the run–time performance can be neglected.

	$COSTRATIO =$		
	5	150	900
Average Deviation	14.42	8.95	7.65
Infeasibility Ratio	10.72	8.12	10.20
Average Run–Time	0.46	0.45	0.44

Table 6.11: The Impact of the Cost Structure on the Performance

The results for different capacity utilizations are provided in Table 6.12. The average deviation from the optimum is positively correlated with the capacity utilization. Differences are, however, not dramatic. Major effects are for the infeasibility ratios. While for each instance with a low capacity utilization a feasible solution can be found, one out of three instances with a capacity utilization $U = 70$ cannot be solved. The run–time performance is almost unaffected by changes in the capacity utilization.

	$U = 30$	$U = 50$	$U = 70$
Average Deviation	9.10	10.99	11.24
Infeasibility Ratio	0.00	2.23	28.84
Average Run–Time	0.46	0.46	0.44

Table 6.12: The Impact of the Capacity Utilization on the Performance

In summary, we find that 100 out of the 1,033 instances in our test–bed cannot be solved by the randomized regret based sampling method. This gives an overall infeasibility ratio of 9.68%. The average run–time is 0.45 CPU–seconds. The overall average deviation of the upper bound from the optimum objective function value is 10.33%.

The most significant method parameter certainly is the number of iterations that are to be performed. Hence, Table 6.13 gives some insight into performance changes due to that value. As a point of reference we also give the results for a variant of the randomized regret based sampling procedures which stops when a first feasible solution is found or when 1,000 iterations are performed.

We see that performing 1,000 iterations is a good choice, because even if we would double that value the average deviation from the optimum objective function values would not be drastically reduced. The poor average deviation results for the first feasible solutions that are found indicates that the instances in the test–bed are indeed not easy to solve. Hence, we can state that the randomized regret based sampling procedure indeed makes a contribution.

	Average Deviation	Infeasibility Ratio	Average Run–Time
First Solution	31.93	9.68	0.02
100 Iterations	13.39	18.01	0.05
500 Iterations	11.00	12.58	0.23
1,000 Iterations	10.33	9.68	0.45
2.000 Iterations	9.78	8.23	0.90

Table 6.13: The Impact of the Number of Iterations on the Performance

6.4 Cellular Automata

The probably most important shortcoming of the randomized regret based sampling method is due to large sizes of the sets $\mathcal{I}_{mt}$ for $m = 1, \ldots, M$ and $t = 1, \ldots, T$. The more elements these sets contain the more biased is the choice of items. In other words, if many items j with a low priority value ϱ_{jt} are contained in a set $\mathcal{I}_{mt}$, but only a very few items with a large priority value are in this set, then it might happen that the probability to choose an item with a low priority value is higher than to choose an item with a high priority value. More formally, let $\mathcal{I}_{mt}^{low}$ and $\mathcal{I}_{mt}^{high}$ be a partition of $\mathcal{I}_{mt}$ for any m and any t with $\mathcal{I}_{mt}$ being a non–trivial set ($| \mathcal{I}_{mt} | \geq 2$), that is,

$$\mathcal{I}_{mt} = \mathcal{I}_{mt}^{low} \cup \mathcal{I}_{mt}^{high} \tag{6.26}$$

and

$$\mathcal{I}_{mt}^{low} \cap \mathcal{I}_{mt}^{high} = \emptyset. \tag{6.27}$$

Consider those partitions only where

$$| \mathcal{I}_{mt}^{low} | \geq | \mathcal{I}_{mt}^{high} | \geq 1 \tag{6.28}$$

and

$$j \in \mathcal{I}_{mt}^{low} \text{ and } i \in \mathcal{I}_{mt}^{high} \Rightarrow \varrho_{jt} \leq \varrho_{it} \tag{6.29}$$

holds. Note, for any set $\mathcal{I}_{mt}$ with $|\mathcal{I}_{mt}| \geq 2$ there are

$$\left\lfloor \frac{|\mathcal{I}_{mt}|}{2} \right\rfloor$$

different partitions fulfilling these assumptions. This expression reveals, the more items are contained in $\mathcal{I}_{mt}$, the more likely it will happen to find a partition with

$$\sum_{j \in \mathcal{I}_{mt}^{low}} \varphi_{mt}(j) > \sum_{j \in \mathcal{I}_{mt}^{high}} \varphi_{mt}(j) \tag{6.30}$$

where the left hand side is the probability to choose an item out of $\mathcal{I}_{mt}^{low}$ and the right hand side is the probability to choose an item out of $\mathcal{I}_{mt}^{high}$. Moreover, in a large set $\mathcal{I}_{mt}$ the maximum sized set $\mathcal{I}_{mt}^{high}$ fulfilling (6.30) tends to include several items. This contradicts the idea of priority rules, because in the usual case the priority values differ decidedly from item to item (otherwise there would not be a need to define complex priority rules since a pure random choice would do roughly the same) and choosing an item with a high priority value should be more probable than choosing an item with a low priority value.

There are at least two ways out of this dilemma. First, which is the most straightforward idea, one could include at most n items into $\mathcal{I}_{mt}$, say $n \leq 5$, which should be those with the highest priority values π_{jt}. Another approach which needs a bit more effort shall be described here. Preliminary studies for the single–machine case are given in [Kim93].

6.4.1 An Introduction to Cellular Automata

Entities in a real–world often do not exist on isolation, but interact with their neighborhood. To study the behavior of a complex system consisting of many entities, researchers thus started to define abstract entities (in contrast to real–world entities), attached attributes to them, and specified simple (deterministic or non–deterministic) interaction effects. Starting with initial attribute values, a new generation of attribute values is then computed by applying the rules of interaction which when done repeatedly simulates the real-world behavior. For example, Conway [Gar70] invented the "Game of Life" to study the life cycles of biological cells. His model is quite simple: Each square on a (very large) chess board represents one cell. By construction, each cell has eight neighboring cells. Attached to each cell is an attribute which indicates either a living or a dead cell. While moving from one generation to the other, a living cell stays alive if and only if it is surrounded by two or three living cells, and a dead cell is revived if and only if it is surrounded by exactly three living cells. Due to this application, such simulation approaches are called cellular automata [ToMa88, Wol86] nowadays.

6.4.2 Masked Setup State Selection

The idea to bring cellular automata and the way setup states are selected together is via a masking mechanism. Suppose that to each pair (j, t) where j is an item index and t is a period index, $mask_{jt}$ is a binary value which is zero if machine m_j must not be set up for item j at the end of period t, and one otherwise. Then, the item sets $\mathcal{I}_{mt}$ can be defined as

$$\mathcal{I}_{mt} \stackrel{def}{=} \{j \in \mathcal{J}_m \mid mask_{jt}(CD_{j(t+1)} + \tilde{d}_{jt}) > 0\} \quad (6.31)$$

$$\cap \{j \in \mathcal{J}_m \mid nr_j - \sum_{\tau=t+1}^{T} q_{j\tau} > 0\}$$

for $m = 1, \ldots, M$ and $t = 1, \ldots, T$. This is to say that an item may only be chosen (using randomized regret based logic) if there is demand for in period t or period $t+1$ and if $mask_{jt} = 1$. Setting some $mask_{jt}$–values to zero thus reduces the set of alternatives for choosing a setup state. Remember, choosing an item randomly will only take place if lots are not split. If $\mathcal{I}_{mt} = \emptyset$, let us choose the dummy item $j_{mt} = 0$.

If we compare this definition of $\mathcal{I}_{mt}$ with (6.10), we can see that items with no demand in period t or period $t+1$ but demand in periods prior to t will not be chosen. However, idle periods are not forbidden which is due to the dummy items. By the way, the construction scheme given in Section 6.2 needs no modification to be applied here.

The concept of cellular automata now comes in as we move from one iteration of the sampling procedure to the next. Starting with $mask_{jt} = 1$ for $j = 1, \ldots, J$ and $t = 1, \ldots, T$ in the first iteration, we simultaneously update all $mask_{jt}$–values when the construction scheme terminates to get a new mask for the next iteration.

In general, fixing the new values $mask_{jt}$ is done stochastically with $prob^1_{jt}$ being the probability the set $mask_{jt} = 1$ and $prob^0_{jt} = 1 - prob^1_{jt}$ being the probability to set $mask_{jt} = 0$. But, there are exceptions where $mask_{jt} = 1$ for some indices j and t. The motivation for these exceptions is that one aspect of the masking mechanism is pushing production amounts into earlier periods in order to bundle demands and thus to build lots. Now, if there is no demand for an item j prior to the period of attention t then it would not make sense to shift future demands into earlier periods. So, we stipulate the following. For $j = 1, \ldots, J$, let

$$t_j \stackrel{def}{=} \min\left\{t \in \{1, \ldots, T\} \mid d^{L4L}_{jt} > 0\right\} \qquad (6.32)$$

be the period which, if a lot–for–lot policy is concerned (see (4.10)), is the latest period so that there is no demand for item j prior to it. Due to the arguments above, for all $j = 1, \ldots, J$ we decide to set $mask_{jt} = 1$ if $t \leq t_j$, and make a random choice for $mask_{jt}$, otherwise.

To give a definition of $prob^1_{jt}$ for $j = 1, \ldots, J$ and $t = t_j + 1, \ldots, T$ that is in accordance with the idea of cellular automata, we need to define some interaction effects on setup states. Two aspects are taken into account here. On the one hand, we consider that each machine can only be in a unique setup state and thus the value

$$S^1_{jt} \stackrel{def}{=} \mid \{i \in \mathcal{J}_{m_j} \mid i \neq j \wedge mask_{it} = 1\} \mid \qquad (6.33)$$

is of special interest. On the other hand, the setup state of a machine should not be changed too often in order to keep setup costs low and thus

$$S_{jt}^2 \stackrel{def}{=} \begin{cases} mask_{j(t-1)} + mask_{j(t+1)} & i, \text{ if } t = t_j + 1, \ldots, T-1 \\ mask_{j(t-1)} & i, \text{ if } t = T \end{cases} \tag{6.34}$$

is also considered. For $k \in \{1, 2\}$ we can discriminate three cases. These are $S_{jt}^k = 0$, $S_{jt}^k = 1$, and $S_{jt}^k \geq 2$. In total there are nine case combinations. We suggest to specify the probability value $prob_{jt}^1$ depending on the case combination that holds (e.g. see Table 6.16).

6.4.3 An Example

Let us consider the data of the example given in Subsection 6.3.5 again. Additionally, suppose the $mask_{jt}$–values as provided in Table 6.14. Note, all entries left to the vertical lines will never change its value while all entries right to the vertical lines are flickering randomly from iteration to iteration.

$mask_{jt}$	$t=1$	2	3	4	5	6	7	8	9	10
$j=1$	1	1	1	1	1	1	1 |	1	0	1
$j=2$	1	1	1	1	1	1 |	0	0	0	0
$j=3$	1	1	1	1	1	1 |	0	1	1	0
$j=4$	1	1	1	1	1 |	1	0	0	1	0

Table 6.14: A Mask for the Selection of Setup States

A protocol of running the construction scheme is provided in Table 6.15. Figure 6.3 shows a plan that could be the outcome when following the lines in the protocol and keep on going until termination.

While the working principle is very much alike the regret based procedure, some interesting points related to the masking mechanism shall be discussed briefly.

Step 1: Item 1 is the only item the machine 1 may be set up for, because there is no other item with demand for in period 10. This is in contrast to the regret based procedure where items

Step	$(t, \Delta t, m)$	$\vec{\tilde{d}}_{jt}$	$\vec{CD}_{j(t+\Delta t)}$	$\mathcal{I}_{mt}$	j_{mt}	$q_{j_{mt}(t+\Delta t)}$
1	(10,1,1)	(20,0,0,0)	(0,0,0,0)	{1}	1	
2	(10,1,2)	(20,0,0,0)	(0,0,0,0)	$\emptyset$	0	
3	(10,0,1)	(20,0,0,0)	(20,0,0,0)			$q_{1T} = 15$
4	(10,0,2)	(20,0,0,0)	(5,0,0,0)			
5	(9,1,1)	(0,15,15,0)	(5,0,0,0)	{1}	1	
6	(9,1,2)	(0,15,15,0)	(5,0,0,0)	{3}	3	$q_{3T} = 0$
7	(9,0,1)	(0,15,15,0)	(5,0,0,0)			$q_{19} = 5$
8	(9,0,2)	(0,15,15,0)	(0,15,15,0)			$q_{39} = 15$
9	(8,1,1)	(0,5,5,20)	(0,15,0,0)	$\emptyset$	0	
10	(8,1,2)	(0,5,5,20)	(0,15,0,0)	{3}	3	
11	(8,0,1)	(0,5,5,20)	(0,15,0,20)			
12	(8,0,2)	(0,5,5,20)	(0,20,5,20)			$q_{38} = 5$
13	(7,1,1)	(20,0,0,5)	(0,20,0,20)	{1}	1	$q_{18} = 0$
14	(7,1,2)	(20,0,0,5)	(0,20,0,20)	$\emptyset$	0	
...						

Table 6.15: A Protocol of the Construction Scheme of the Cellular Automaton Based Procedure

with demand in periods prior to the current period could also be selected.

Step 2: In period 10 is no demand for any item that may be produced on machine 2. Hence, the machine is set up for the dummy item 0 (which will be corrected by the postprocessor).

Step 5: Despite $mask_{19} = 0$, we choose item 1 again to avoid lot splitting.

Step 6: Since $mask_{29} = 0$, item 3 is the only item machine 2 may be set up for at the end of period 9.

Step 9/13: Due to $mask_{48} = mask_{47} = 0$, machine 1 is kept idle during period 8.

Step 14: Again, machine 2 is set up for the dummy item, because the mask prohibits to set the machine up for item 2 (though there is demand for item 2).

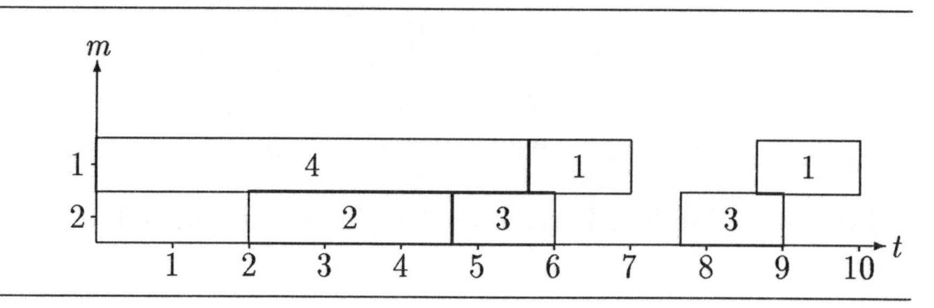

Figure 6.3: A Possible Outcome of the Run in the Protocol

6.4.4 Experimental Evaluation

To study the performance of the cellular automaton based heuristic, we use the same method parameter settings as in Subsection 6.3.6. Especially the number of iterations is chosen to be 1,000 again to allow fair comparisons. The probabilities $prob^1_{jt}$ are defined on the basis of the parameter values given in Table 6.16. To decide whether a certain mask entry is set to one or not, we draw two random numbers out of the interval $[0, 1]$ with uniform distribution and multiply them. If the result is less than or equal to the number in Table 6.16 that corresponds to the case combination that holds, the mask entry is set to one, otherwise it is set to zero. The reason for this strategy comes from the believe that it is easier to specify a probability to set a mask entry to one for each case on isolation. The data in Table 6.16 are generated as follows: If the case $S^1_{jt} = 0$ holds we would like to have a probability of 0.8 for a mask entry to be set to one (analogously, we define 0.5 for the case $S^1_{jt} = 1$, and 0.2 for the case $S^1_{jt} \geq 2$). Likewise, we assume 0.2 if $S^2_{jt} = 0$ holds (and 0.5 for $S^2_{jt} = 1$ and 0.8 for $S^2_{jt} = 2$). For the case combination $S^1_{jt} = 0$ and $S^2_{jt} = 0$, for example, we then derive $0.8 \cdot 0.2 = 0.16$ as given in the table. Make sure to understand that this is not the probability that the case combination under concern will hold.

The details of the computational study for the small PLSP–MM–instances which are defined in Section 4.2 are given in the Tables B.4 to B.6 in the Appendix B. Again, these tables show for

	$S^2_{jt} = 0$	$S^2_{jt} = 1$	$S^2_{jt} = 2$
$S^1_{jt} = 0$	0.16	0.40	0.64
$S^1_{jt} = 1$	0.10	0.25	0.40
$S^1_{jt} \geq 2$	0.04	0.10	0.16

Table 6.16: Method Parameters for Computing Mask Entries

each parameter level combination the average deviation from the optimum objective function value, the worst case deviation, and the number of instances for which a feasible solution is found. Formula (6.25) is used to compute the deviation of the upper bound from the optimum result.

The effects of changes in the parameter levels are investigated on the basis of aggregated data here. Table 6.17 reveals what the number of machines affects. Every performance measure is positively correlated with the number of machines. The average deviation is almost doubled when instances with two instead of one machine are solved.

	$M = 1$	$M = 2$
Average Deviation	7.66	14.29
Infeasibility Ratio	11.84	14.09
Average Run–Time	0.48	0.60

Table 6.17: The Impact of the Number of Machines on the Performance

The gozinto–structure complexity is examined in Table 6.18. It turns out that its impact on the average deviation from optimum and on the run–time performance is small. The infeasibility ratio, however, is decidedly larger when instances with a high complexity are solved.

Table 6.19 shows what different demand patterns do cause. For the average deviation from the optimum results there is a clear

	$\mathcal{C} = 0.2$	$\mathcal{C} = 0.8$
Average Deviation	10.88	11.01
Infeasibility Ratio	10.48	15.55
Average Run–Time	0.53	0.55

Table 6.18: The Impact of the Gozinto–Structure Complexity on the Performance

tendency to be large when instances with many non–zeroes in the demand matrix are solved, and to be small if the demand matrices are sparsely–filled. For the infeasibility ratio, however, the trend is not the same. Though sparsely–filled demand matrices do give much better results, too, more entries in the demand matrix do not necessarily mean a worse result. The run–time increases with the number of mask entries that are to be updated within each iteration.

	$(T_{macro}, T_{micro}, T_{idle}) =$		
	$(10, 1, 5)$	$(5, 2, 2)$	$(1, 10, 0)$
Average Deviation	17.50	13.39	3.87
Infeasibility Ratio	17.66	21.60	0.56
Average Run–Time	0.60	0.57	0.48

Table 6.19: The Impact of the Demand Pattern on the Performance

The impact of the cost structure on the performance is studied in Table 6.20. It can be seen that large setup costs result in a dramatically better average deviation from the optimum result than low setup costs. However, the infeasibility ratio increases when the ratio of setup and holding costs does. The run–time performance is totally unaffected.

Performance is also affected by the capacity utilization as one can see in Table 6.21. The average deviation from the optimum objective function value grows with the capacity utilization. For high

	$COSTRATIO =$		
	5	150	900
Average Deviation	18.94	8.85	4.91
Infeasibility Ratio	12.46	13.04	13.41
Average Run–Time	0.54	0.54	0.54

Table 6.20: The Impact of the Cost Structure on the Performance

capacity utilization, the infeasibility ratio indicates very poor performance. The reason why run–time performance is better for high utilization than for low may be explained by that fact. If capacity usage is high, it is hard to find a feasible solution. Hence, many iterations are terminated after a feasibility check which reduces the average run–time.

	$U = 30$	$U = 50$	$U = 70$
Average Deviation	8.31	11.97	13.94
Infeasibility Ratio	0.00	1.96	39.81
Average Run–Time	0.56	0.55	0.49

Table 6.21: The Impact of the Capacity Utilization on the Performance

In summary, we have that 134 instances out of the 1,033 instances in the test–bed cannot be solved with the cellular automaton based method. This is an overall infeasibility ratio of 12.97%. The average run–time is 0.54 CPU–seconds. The average deviation from the optimum objective function value is 10.94%.

To see if the masking mechanism of the cellular automaton makes a significant contribution, we compare the results of the heuristic as described above with the results that come out when no masking helps to select items. Table 6.22 shows the data.

While the infeasibility ratio is almost unaffected by the masking mechanism, the average deviation from the optimum is deci-

	Average Deviation	Infeasibility Ratio	Average Run–Time
With Masking	10.94	12.97	0.54
Without Masking	17.21	12.39	0.41

Table 6.22: The Impact of the Masking Mechanism on the Performance

dedly better if masking supports the choice of setup states. The price for the performance improvement of the average deviation are larger run–times.

6.5 Genetic Algorithms

A key element of what is assumed to be intelligence is the capability to learn from past experience. Especially when things are done repeatedly, intelligent behavior would avoid doing a mistake more than once and would prefer making advantageous decisions again.

None of the methods introduced thus far utilizes learning effects (beside in the tuning of methods parameters as explained in Subsection 6.3.3). Hence, we will present a method in this section that features learning.

6.5.1 An Introduction to Genetic Algorithms

For optimization a class of today's most popular heuristic approaches is known as genetic algorithms. Due to its widespread use and the vast amount of literature dealing with genetic algorithms, e.g. [Dav91, Gol89, Hol75, LiHi89, MüGoKr88, Ree93, Whi93], a comprehensive review of research activities is doomed to failure. Thus, we stick to an outline of the fundamental ideas.

The adjective genetic reveals the roots of these algorithms. Adapting the evolution strategy from natural life forms, the basic idea is to start with a set of (feasible) solutions and to compute a set of new solutions by applying some well–defined operators on the old ones. Then, some solutions (new and/or old ones) are selected to form a new set with which another iteration is started, and so on until some stopping criterion is met. Solutions are represented by sets of attributes, and different solutions are represented by different collections of attribute values. The decision which solutions are dismissed and which are taken over to form a new starting point for the next iteration is made on the basis of a priority rule.

Most authors use notions from evolution theory in this context. The set of solutions an iteration starts with is usually called the parent population while the set of new solutions is the child population. Each iteration represents a generation. A member of a population is an individual or a chromosome, thus we have parent and child individuals (or chromosomes). The attribute values that belong to an individual are called genes. This coins the name of this type of algorithm. The operations for procreating new individuals are applications of so–called genetic operators. Attached to each individual is a fitness value which functions as a priority rule to select the parent individuals for the next generation. This mechanism should simulate what is observed in nature where only the fittest survive and the weak die, good characteristics are inherited and bad ones become extinct.

Up to here, there are many degrees of freedom and thus genetic algorithms are often called meta–heuristics. To develop a method based on the ideas of genetic algorithms for a specific problem, we need to provide some more ingredients. First, we need to specify how to encode a solution of the problem as a set of attributes. Furthermore, a definition of how to compute fitness values needs to be given. Also, we need to define genetic operators. Eventually, the way to select a new parent population must be described. Of minor importance, but not to forget, is a stopping criterion, e.g. a total number of iterations, set by the user. The population sizes denoted as $PARENT$ for the size of the parent population and $CHILD$ for the number of child individuals, respectively, are specified by the user, too.

6.5.2 Problem Representation

Traditionally, a solution of a given problem is represented as a bitstring, i.e. a sequence of binary values [Gol89]. For many problems this gives not a very compact representation of solutions. So, in some applications genes are chosen to be more complex rather than being binary–valued only. See for instance [DoPe95] for an application to job shop scheduling where a gene represents a rule to select a job for scheduling.

For representing a solution of the PLSP, we have chosen a two–dimensional matrix with M rows and T columns. Since we consider a population of matrices, let each matrix be identified by a unique label

$$k \in \{1, \ldots, PARENT, PARENT+1, \ldots, PARENT+CHILD\}. \tag{6.35}$$

An entry in row m and column t in the matrix k is a rule $\vartheta_{mtk} \in \Theta$ for selecting the setup state for machine m at the end of period t out of the set $\mathcal{I}_{mt}$. Here, Θ denotes the set of all selection rules which is to be defined. To get things straight, recall that the matrices are the individuals now, and that selection rules are genes.

6.5.3 Setup State Selection Rules

Though the rules to select setup states is a detail that can be skipped on first reading, it certainly is a significant aspect for the performance of the construction scheme. We will now suggest several rules for selecting a setup state for machine m at the end of period t where $m = 1, \ldots, M$ and $t = 1, \ldots, T$. In the following, let us assume that whenever ties are to be broken, we favor items with a low index by arbitration. If $\mathcal{I}_{mt} = \emptyset$ given the definitions below, then we choose the dummy item $j_{mt} = 0$.

Rule θ_1: Maximum Holding Costs

Consider those items for which there is demand in period $t+1$, i.e.

$$\mathcal{I}_{mt} \stackrel{def}{=} \{j \in \mathcal{J}_m \mid CD_{j(t+1)} > 0\} \cap \{j \in \mathcal{J}_m \mid nr_j - \sum_{\tau=t+1}^{T} q_{j\tau} > 0\}. \tag{6.36}$$

When setting machine m up for an item i,

$$\sum_{j \in \mathcal{I}_{mt} \setminus \{i\}} h_j CD_{j(t+1)}$$

are the holding costs that are charged to keep the remaining items in inventory. Note, this is just an estimate which assumes that all $CD_{i(t+1)}$ items can indeed be scheduled in period $t+1$. If this is not true, item i would incur holding costs, too. To keep these costs low we should choose an item causing high holding costs, i.e.

$$j_{mt} \in \left\{ i \in \mathcal{I}_{mt} \mid h_i CD_{i(t+1)} = \max_{j \in \mathcal{I}_{mt}} \{h_j CD_{j(t+1)}\} \right\}. \tag{6.37}$$

Rule θ_2: Minimum Setup Costs

In order to keep setup costs low, we choose

$$j_{mt} = j_{m(t+1)}, \tag{6.38}$$

if $j_{m(t+1)} \neq 0$ and $\tilde{d}_{j_{m(t+1)}t} > 0$. If this does not hold, we consider the items with demand in period t or period $t+1$, i.e.

$$\mathcal{I}_{mt} \stackrel{def}{=} \{j \in \mathcal{J}_m \mid CD_{j(t+1)} + \tilde{d}_{jt} > 0\} \cap \{j \in \mathcal{J}_m \mid nr_j - \sum_{\tau=t+1}^{T} q_{j\tau} > 0\}, \tag{6.39}$$

and choose the one with lowest setup costs. That is,

$$j_{mt} \in \left\{ i \in \mathcal{I}_{mt} \mid s_i = \min_{j \in \mathcal{I}_{mt}} \{s_j\} \right\}. \tag{6.40}$$

Rule θ_3: Introduce Idle Periods

To enforce keeping a machine idle we allow to choose items for which there is demand in periods prior to t. In this case,

$$\mathcal{I}_{mt} \stackrel{def}{=} \{j \in \mathcal{J}_m \mid CD_{j(t+1)} + \sum_{\tau=1}^{t} \tilde{d}_{j\tau} > 0\} \tag{6.41}$$

$$\cap \{j \in \mathcal{J}_m \mid nr_j - \sum_{\tau=t+1}^{T} q_{j\tau} > 0\}$$

is the item set under consideration. For $j \in \mathcal{I}_{mt}$ we determine

$$t_j \stackrel{def}{=} \begin{cases} t+1 & \text{, if } CD_{j(t+1)} > 0 \\ \max\{\tau \mid 1 \leq \tau \leq t \wedge \tilde{d}_{j\tau} > 0\} & \text{, otherwise} \end{cases} \tag{6.42}$$

which is the latest period less than or equal to $t+1$ with demand for item j. Since idle periods bear the risk to lead to an infeasible result, idle periods should not last too long. Hence, we choose

$$j_{mt} \in \left\{ i \in \mathcal{I}_{mt} \mid t_i = \max_{j \in \mathcal{I}_{mt}} \{t_j\} \right\}. \tag{6.43}$$

Note, updating the demand matrix as done in Table 6.5 is not necessary here, because

$$\sum_{\tau=t_{j_{mt}}+1}^{t} \tilde{d}_{j\tau} = 0$$

holds for all items $j \in \mathcal{I}_{mt}$ by construction.

Rule θ_4: Maximum Depth

To avoid infeasibility, it might be a good idea to choose items with a large depth. Thus, taking the items given by (6.39) into account, the setup state should be chosen using

$$j_{mt} \in \left\{ i \in \mathcal{I}_{mt} \mid dep_i = \max_{j \in \mathcal{I}_{mt}} \{dep_j\} \right\}. \tag{6.44}$$

Rule θ_5: Maximum Number of Predecessors

Quite similar to rule θ_4 is the rule proposed now. This time, we take the total number of predecessors into account. Again, consider the items defined by (6.39) and choose

$$j_{mt} \in \left\{ i \in \mathcal{I}_{mt} \,\middle|\, |\bar{\mathcal{P}}_i| = \max_{j \in \mathcal{I}_{mt}} \{|\bar{\mathcal{P}}_j|\} \right\}. \tag{6.45}$$

Rule θ_6: Maximum Demand for Capacity

Determining the capacity utilization of the bottleneck machine also tends to avoid infeasible solutions. Focusing on the items defined in (6.39) again, we compute

$$cap_j \stackrel{def}{=} (CD_{j(t+1)} + \tilde{d}_{jt}) \cdot \max\left\{ \frac{\sum_{i \in (\mathcal{P}_j \cup \{j\}) \cap \mathcal{J}_m} p_i i d_{ji}}{\sum_{\tau=1}^{t} C_{m\tau}} \mid m \in \{1, \ldots, M\} \right\} \tag{6.46}$$

for $j \in \mathcal{I}_{mt}$ (compare (6.14)). Afterwards, we choose

$$j_{mt} \in \left\{ i \in \mathcal{I}_{mt} \mid cap_i = \max_{j \in \mathcal{I}_{mt}} \{cap_j\} \right\}. \tag{6.47}$$

Rule θ_7: Pure Random Choice

Last, j_{mt} can be chosen out of the set given by (6.39) with a pure random choice to give items with no extreme characteristic a chance to be selected.

In summary, we have

$$\Theta \stackrel{def}{=} \{\theta_1, \theta_2, \theta_3, \theta_4, \theta_5, \theta_6, \theta_7\} \tag{6.48}$$

and sets of items $\mathcal{I}_{mt}$ to choose among as defined above. This is what is used in our tests. Note, following our arguments for choosing composite priority values for the regret based method in Section 6.3, we have introduced both, rules that tend to give cheap production plans and rules that tend to give feasible plans. In contrast to a composite criterion, the rules given here need less effort to be evaluated. All rules but θ_7 operate deterministically.

6.5.4 Fitness Values

To compute a fitness value $fitness_k$ for an individual k we call the construction scheme using the selection rules $\vartheta_{mtk} \in \Theta$ for choosing the setup states. Let $fitness_k$ be the objective function value of the production plan that is constructed when matrix k is used (and let $fitness_k = \infty$, if no feasible plan was found using matrix k). It should be clear that due to this definition searching for an individual with utmost fitness in fact means to look for an individual with lowest possible fitness value.

6.5.5 Genetic Operators

In order to generate a new parent population out of an old one, we employ three different operators. First, a so–called crossover combines two parent individuals to procreate one child individual. Second, mutation introduces non–determinism into the inheritance. And third, a selection filters the new parent population out of the last generation. The details of these operators shall be given now.

The crossover operates on two matrices, say k_1 and k_2 where $k_1, k_2 \in \{1, \ldots, PARENT\}$. Applying a crossover then cuts the two matrices into four pieces each and puts some of the submatrices together yielding a new matrix $k_3 \in \{PARENT + 1, \ldots, PARENT + CHILD\}$ of the same size. For doing so, suppose that two numbers $\hat{m}_{k_3} \in \{1, \ldots, M\}$ and $\hat{t}_{k_3} \in \{1, \ldots, T\}$ are given. More formally, the resulting matrix k_3 is defined as

$$\vartheta_{mtk_3} \stackrel{def}{=} \begin{cases} \vartheta_{mtk_1} & \text{, if } m \leq \hat{m}_{k_3} \text{ and } t \leq \hat{t}_{k_3} \\ \vartheta_{mtk_2} & \text{, if } m \leq \hat{m}_{k_3} \text{ and } t > \hat{t}_{k_3} \\ \vartheta_{mtk_2} & \text{, if } m > \hat{m}_{k_3} \text{ and } t \leq \hat{t}_{k_3} \\ \vartheta_{mtk_1} & \text{, if } m > \hat{m}_{k_3} \text{ and } t > \hat{t}_{k_3} \end{cases} \tag{6.49}$$

for $m = 1, \ldots, M$ and $t = 1, \ldots, T$.

The mutation stochastically changes some entries of a matrix k. Let $MUTATION \in [0, 1]$ be a (small) probability to change an entry. Furthermore, suppose $prob_{mt} \in [0, 1]$ is drawn at random with uniform distribution where $m = 1, \ldots, M$ and $t = 1, \ldots, T$. Then, the mutation of matrix k is defined as

$$\vartheta'_{mtk} = \begin{cases} \vartheta_{mtk} & \text{, if } prob_{mt} \geq MUTATION \\ \hat{\theta}_{mt} & \text{, otherwise} \end{cases} \tag{6.50}$$

for $m = 1, \ldots, M$ and $t = 1, \ldots, T$, where $\hat{\theta}_{mt}$ is drawn at random out of Θ with uniform distribution.

The selection of $PARENT$ individuals which form a new parent population is done deterministically choosing those matrices with the highest fitness values. Ties are broken randomly. The effort to find these is the effort of sorting $PARENT + CHILD$ objects. Without loss of generality, we assume the selected individuals be relabeled having unique indices $k = 1, \ldots, PARENT$.

6.5.6 The Working Principle in a Nutshell

Initially, the genetic algorithm starts with a parent population that is randomly generated by drawing a rule ϑ_{mtk} for each position (m,t) in the matrix k with uniform distribution out of the set of rules Θ where $m = 1,\ldots,M$, $t = 1,\ldots,T$, and $k = 1,\ldots,PARENT$. Then, we compute the fitness values for the matrices $k = 1,\ldots,PARENT$. To do so, we have to execute the construction scheme a total of $PARENT$ times.

Afterwards, a population of $CHILD$ individuals with unique indices $k = PARENT + 1,\ldots,PARENT + CHILD$ is generated using the crossover operation. The two parent individuals that are combined to create a new child individual k are randomly chosen out of $\{1,\ldots,PARENT\}$ with uniform distribution. The values $\hat{m}_k$ and $\hat{t}_k$ used as parameters for the crossover operators are integral random numbers which are drawn out of $[1,\ldots,M]$ and $[1,\ldots,T]$, respectively, with uniform distribution. Mutation of all child individuals is done next. Eventually, the fitness values for the matrices $k = PARENT + 1,\ldots,PARENT + CHILD$ are computed executing the construction scheme $CHILD$ times. Finally, the parent population for the next generation is selected having (new) indices $k = 1,\ldots,PARENT$. The process is repeated starting with the generation of new child individuals until some stopping criterion is met.

The production plan with the lowest objective function value found during all iterations is given as a result.

6.5.7 An Example

Once more we use the data of the example given in Subsection 6.3.5. Let us suppose that $s_2 < s_3$ holds. Furthermore, assume a matrix k filled with selection rules as given in Table 6.23.

A protocol of running the construction scheme is shown in Table 6.24. Figure 6.4 depicts a plan that could be the outcome when completing the protocol.

Some interesting points shall be explained in a little more detail.

ϑ_{mtk}	$t=1$	2	3	4	5	6	7	8	9	10
$m=1$	θ_7	θ_4	θ_1	θ_7	θ_3	θ_5	θ_6	θ_3	θ_5	θ_4
$m=2$	θ_2	θ_6	θ_4	θ_1	θ_7	θ_3	θ_5	θ_1	θ_2	θ_7

Table 6.23: A Matrix of Setup State Selection Rules

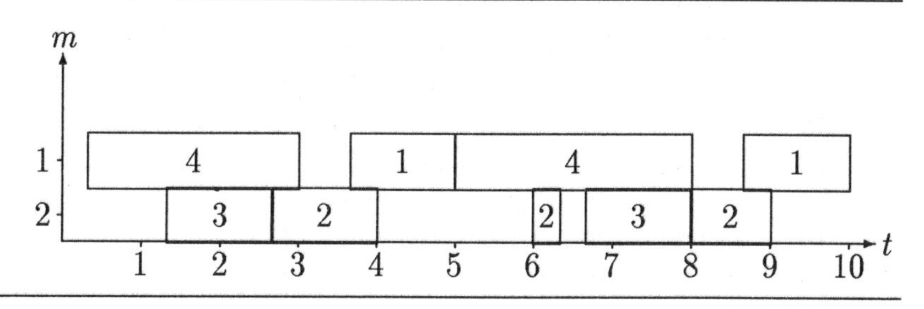

Figure 6.4: A Possible Outcome of the Run in the Protocol

Step 2: The set of items $\mathcal{I}_{2T}$ to choose among is empty. Hence, we choose the dummy item 0.

Step 5: Item 1 is chosen again, because lot splitting is not allowed.

Step 6: Item 2 is chosen due to the selection rule $\vartheta_{29k} = \theta_2$ which chooses the item with the lowest setup costs. Remember, we have assumed $s_2 < s_3$.

Step 9: To set machine 1 up at the end of period 8, we use the selection rule $\vartheta_{18k} = \theta_3$ which may introduce idle periods. In the item set $\mathcal{I}_{18}$ we have both, item 1 and item 4, because there is demand for item 1 in period 7 and demand for item 4 in period 8. Since idle time is kept as short as possible, item 4 is chosen.

Step 10: The selection rule to be employed is $\vartheta_{28k} = \theta_1$ which chooses the item with the maximum holding costs. For item 3 being the only one with cumulative demand, $\mathcal{I}_{28} = \{3\}$ and no other item is contained in the item set. As one can see, the capacity of machine 2 in period 9 is used up by item 2 which was scheduled in Step 8. Thus, item 3 cannot be scheduled in period 9, but in

Step	$(t, \Delta t, m)$	$\vec{\tilde{d}}_{jt}$	$\vec{CD}_{j(t+\Delta t)}$	$\mathcal{I}_{mt}$	j_{mt}	$q_{j_{mt}(t+\Delta t)}$
1	(10,1,1)	(20,0,0,0)	(0,0,0,0)	$\{1\}$	1	
2	(10,1,2)	(20,0,0,0)	(0,0,0,0)	$\emptyset$	0	
3	(10,0,1)	(20,0,0,0)	(20,0,0,0)			$q_{1T} = 15$
4	(10,0,2)	(20,0,0,0)	(5,0,0,0)			
5	(9,1,1)	(0,15,15,0)	(5,0,0,0)	$\{1\}$	1	
6	(9,1,2)	(0,15,15,0)	(5,0,0,0)	$\{2,3\}$	2	$q_{2T} = 0$
7	(9,0,1)	(0,15,15,0)	(5,0,0,0)			$q_{19} = 5$
8	(9,0,2)	(0,15,15,0)	(0,15,15,0)			$q_{29} = 15$
9	(8,1,1)	(0,5,5,20)	(0,0,15,0)	$\{1,4\}$	4	$q_{49} = 0$
10	(8,1,2)	(0,5,5,20)	(0,0,15,0)	$\{3\}$	3	$q_{39} = 0$
11	(8,0,1)	(0,5,5,20)	(0,0,15,20)			$q_{48} = 15$
12	(8,0,2)	(0,5,5,20)	(0,5,20,5)			$q_{38} = 15$
13	(7,1,1)	(20,0,0,15)	(0,5,5,5)	$\{4\}$	4	
14	(7,1,2)	(20,0,0,15)	(0,5,5,5)	$\{3\}$	3	
...						

Table 6.24: A Protocol of the Construction Scheme of the Genetic Algorithm

period 8 (Step 12).

6.5.8 Experimental Evaluation

To test the performance, the genetic algorithm is applied to the small PLSP–MM–instances which are defined in Section 4.2. The mutation probability is chosen to be 0.1. For allowing fair comparisons with other heuristics, the genetic algorithm should not perform more executions of the construction phase than the others do. Hence, the method parameters are chosen as $PARENT = 20$, $CHILD = 10$, and 98 being the number of generations. This gives a total of 1,000 runs of the construction phase. Tables B.7 to B.9 in the Appendix B provide detailed results in the usual format, i.e. for each parameter level they show the average deviation from the optimum objective function value, the worst case deviation, and the number of instances for which a feasible solution is found.

To find out which parameter levels have what effect on the performance, we aggregate the data. Table 6.25 focuses on the number of machines. As we can see, additional machines increase the average deviation from the optimum, but reduce the infeasibility ratio. In both cases, the effect is remarkably large. Only small changes are measured for the run–time performance.

	$M = 1$	$M = 2$
Average Deviation	17.72	21.90
Infeasibility Ratio	20.39	14.67
Average Run–Time	0.08	0.11

Table 6.25: The Impact of the Number of Machines on the Performance

Table 6.26 examines the impact of the gozinto–structure complexity on the performance. It becomes clear that a high complexity has a drastic negative effect on both, the average deviation from the optimum as well as the infeasibility ratio. The run–time performance is not affected.

	$\mathcal{C} = 0.2$	$\mathcal{C} = 0.8$
Average Deviation	18.21	21.80
Infeasibility Ratio	13.71	21.46
Average Run–Time	0.10	0.10

Table 6.26: The Impact of the Gozinto–Structure Complexity on the Performance

For different demand patterns, Table 6.27 shows that many positive entries in the demand matrix have a dramatic effect on the average deviation from the optimum and on the infeasibility ratio. The results turn out to be very poor. The run–time performance, however, does not change.

	$(T_{macro}, T_{micro}, T_{idle}) =$		
	$(10,1,5)$	$(5,2,2)$	$(1,10,0)$
Average Deviation	36.68	22.19	6.69
Infeasibility Ratio	32.19	18.21	2.51
Average Run–Time	0.10	0.10	0.10

Table 6.27: The Impact of the Demand Pattern on the Performance

An investigation of different cost structures is performed in Table 6.28. Clearly, this parameter level has a significant impact on the average deviation from the optimum. While low setup costs give the worse result, high setup costs give only second best results. The best average deviation is reached for a balanced cost structure. The infeasibility ratio and the run–time performance are almost unaffected by different costs.

	$COSTRATIO =$		
	5	150	900
Average Deviation	26.20	15.78	17.66
Infeasibility Ratio	17.39	17.68	17.49
Average Run–Time	0.10	0.10	0.10

Table 6.28: The Impact of the Cost Structure on the Performance

The capacity utilization is studied in Table 6.29. The best average deviation from the optimum is gained for a high utilization. However, the infeasibility ratio grows quickly when the capacity utilization is increased. For $U = 70$, four out of ten instances cannot be solved. Once more, the run–time performance remains stable.

In summary, the genetic algorithm is unable to solve 181 out of the 1,033 instances in the test–bed. This corresponds to an overall infeasibility ratio of 17.52%. The average run–time is 0.10 CPU–seconds. The average deviation from the optimum objective

	$U = 30$	$U = 50$	$U = 70$
Average Deviation	19.75	21.41	17.60
Infeasibility Ratio	0.00	13.13	42.01
Average Run–Time	0.10	0.10	0.10

Table 6.29: The Impact of the Capacity Utilization on the Performance

function value is 19.89%.

The most important method parameters of the genetic algorithm are the sizes of the parent and the child population. Hence, Table 6.30 gives some insight into what happens if these values are varied. All other parameters are kept as they are.

$PARENT$	$CHILD$	Average Deviation	Infeasibility Ratio	Average Run–Time
20	10	19.89	17.52	0.10
200	10	113.73	0.87	0.26
200	100	14.75	8.71	1.10

Table 6.30: The Impact of the Population Sizes on the Performance

It turns out that increasing the population sizes reduces the infeasibility ratio remarkably. Only nine out of 1,033 instances are left unresolved when we choose $PARENT = 200$ and $CHILD = 10$. With respect to the average deviation from the optimum, it becomes clear that the parent population should not be chosen too large in comparison with the child population. Since the genetic algorithm works very fast, it is no problem to evaluate a large number of calls to the construction scheme. For 98 generations where the parent population contains 200 individuals and each child population contains 100 individuals we have to execute the construction scheme 10,000 times which can be done in round about one second.

6.6 Disjunctive Arc Based Tabu Search

In this section we combine both, a graphical representation of PLSP–instances and a modern local search method operating on these graphs. Together with a modification of the construction scheme which uses such a graph structure as input, we have a complex method to attack the PLSP.

Preliminary tests for the multi–level, single–machine PLSP using tabu search strategies are given in [Kim94a].

6.6.1 An Introduction to Tabu Search

Tabu search approaches performed well on a wide variety of optimization problems and its ideas can frequently be found in articles that give recent benchmark results. Reports on these sophisticated local search procedures are given in an uncountable number of publications. We refer to [FaKe92, Glo89, Glo90a, Glo90b, Glo94, HeWe91] and restrict ourself to a brief review of the underlying concepts.

Starting with an initial solution, the idea of local search (as considered here) is to apply some well–defined operations to it and, by doing so, to move to a new (feasible) solution which is used as a starting point for the next iteration. Since there may be a choice of what operations to apply to move from one solution to another, we usually have a neighborhood consisting of more than one solution that can be reached starting with a particular solution and performing one step. Since we face an optimization problem, one should choose those operations which lead to a neighbor with best, say lowest, objective function value of all solutions in the neighborhood. This does not guarantee to decrease the objective function values monotonically, and thus the strategy is sometimes also referred to as steepest descent, mildest ascent [Glo89, HaJa90]. The characteristic feature of a tabu search method is the existence of a so–called tabu list which contains the reverse operations of some recently executed operations. Operations may only be performed if they are not included in the tabu list. Depending on the size of the list, this mechanism avoids cycling and helps to escape from local optima. However, in some situations one might wish to execute an operation although it is tabu. Hence, a so–called aspiration criterion is used to decide whether or not an operation is indeed tabu.

The tabu search idea again is a meta–heuristic, and many details are left unspecified. To tackle a specific problem such as the PLSP with a tabu search method, we need to introduce a representation of solutions, operations that can be applied to these representations, a definition of the tabu list, and aspiration criteria. All that is given in the subsections below.

6.6.2 The Data Structure

First of all, we need an appropriate representation of the solution space to search in. A first hint is given by the graphical representation of the gozinto–structure as a gozinto–tree (see Subsection 3.3.3). For each positive entry d_{jt} in the external demand matrix the internal demands must be fulfilled respecting the precedence relations in the gozinto–tree Γ_j.

The underlying idea of the method described in this section is to introduce additional non–redundant arcs into the gozinto–trees. These arcs represent precedence relations, too, and thus affect the construction of a (feasible) solution (see also [Bal69] for the concept of so–called disjunctive arcs). A tabu search is performed to find an orientation of the disjunctive arcs that gives a production plan with a low objective function value. Roughly speaking, the procedure developed here works as follows: Starting with an initial graphical representation of the PLSP–instance at hand which consists of a gozinto–tree for each positive external demand entry plus some disjunctive arcs, a first production plan is constructed respecting the precedence relations in that graph. Then, the orientation of some arcs is changed and a next plan is generated. This goes on until some stopping criterion is met while a tabu search guides the process of reorienting arcs.

It should be remarked that the proposed tabu search strategy does not strictly move from one feasible solution to another. Such a strategy could be termed "tunneling" to express the hope that while bypassing infeasible solutions a feasible solution can be found again.

The data structure Γ^{TS} on which the procedure operates on can formally be defined as[7]

$$\Gamma^{TS} \stackrel{def}{=} (\bigcup_{j=1}^{J} \bigcup_{t=1}^{T} \tilde{\Gamma}_j^{TS}(t, d_{jt}, \omega_1(j,t)), \mathcal{DA}) \tag{6.51}$$

where $\tilde{\Gamma}_j^{TS}$ is a function similar to (3.21) which generates a gozinto–tree with item j being its root node plus some disjunctive arcs inside this tree. The function ω_1 will be used to assign unique labels to every node. $\mathcal{DA}$ is the set of disjunctive arcs.

$$\tilde{\Gamma}_j^{TS}(period, demand, label) \stackrel{def}{=} \begin{cases} \emptyset & \text{, if } demand = 0 \\ \{ (j, \\ \quad period, \\ \quad period, \\ \quad demand, \\ \quad label, \\ \quad \bigcup_{i \in \mathcal{P}_j} \tilde{\Gamma}_i^{TS} \; (period - v_i, \\ \qquad\qquad a_{ij} demand, \\ \qquad\qquad \omega_2(j, i, label)))\} \\ & \text{, otherwise} \end{cases} \tag{6.52}$$

where ω_2 is a function which will be used to define unique labels for internal demand nodes.

In contrast to the definition of a gozinto–tree as given in (3.21) each node is now represented as a six–tuple. Again, the first entry is the number of the respective item and the last entry is the set of immediately preceding nodes. At the second position we find the deadline until which the demand represented by the node is to be met. This deadline is known for external demands only, hence we temporarily fill in a deadline derived from a lot–for–lot policy at this position for internal demands and update the entry at this position as we proceed. The third position also contains a period's index. It is the deadline of the node when scheduling is done on a

[7]The upper index TS stands for Tabu Search.

lot–for–lot basis. Initially, the entries at position two and three are equal. At position four we have the demand size that is represented by the node. For root nodes this equals the external demand. The fourth position contains a unique label to identify each node.

What is left now, is a specification of how to compute unique labels by means of ω_1 and ω_2, and how to introduce disjunctive arcs. Let us start with the discussion of unique labels first. To begin with, suppose that we have to compute unique labels for the nodes of a single gozinto–tree. Let $nodes_j$ denote the number of nodes in a gozinto–tree with item j being its root node where $j = 1, \ldots, J$. It can recursively be computed using

$$nodes_j \stackrel{def}{=} \sum_{i=1}^{J} nodes_{ji} \tag{6.53}$$

where

$$nodes_{ji} \stackrel{def}{=} \begin{cases} 1 & \text{, if } j = i \\ 0 & \text{, if } j \neq i \text{ and } i \notin \bar{\mathcal{P}}_j \\ \sum_{h \in \mathcal{S}_i} nodes_{jh} & \text{, otherwise} \end{cases} \tag{6.54}$$

is the number of nodes with item index i in the gozinto–tree having item j as its root node. Now, let us assume that we already have a label for the root node of a gozinto–tree. Then, we can recursively traverse the gozinto–tree level–by–level assigning unique labels to each node in the tree. More formally, let i be the item number of a node under consideration in the tree and j be the item number of the unique successor node. Furthermore, let $label$ be the unique label of the successor node which exists by construction. Then,

$$\omega_2(j, i, label) \stackrel{def}{=} 1 + label + \sum_{\substack{k \in \mathcal{P}_j \\ k < i}} nodes_k \tag{6.55}$$

is a unique label for the node under consideration. What remains to do is to give a definition of root node labels which guarantee that nodes in different gozinto–trees do not have identical labels. The function

$$\omega_1(j, t) \stackrel{def}{=} \sum_{i=1}^{J} \sum_{\tau=1}^{t-1} \chi_2(d_{i\tau}) nodes_i + \sum_{i=1}^{j-1} \chi_2(d_{it}) nodes_i \tag{6.56}$$

determines the label for the root node that corresponds to a positive entry for item j in period t in the external demand matrix where χ_2 is an auxiliary function defined in (3.40). In summary, we have uniquely labeled all nodes in the graphical representation with integral numbers out of the interval

$$\mathcal{L} \stackrel{def}{=} [0, \ldots, \sum_{j=1}^{J} \sum_{t=1}^{T} \chi_2(d_{jt}) nodes_j - 1] \tag{6.57}$$

which, by the way, can neatly be used for accessing nodes in a computer implementation.

Let us now turn to the disjunctive arcs that define additional precedence relations among the nodes. It should be carefully considered what arcs to integrate. This is not only to avoid redundant arcs, but also to avoid precedence relations that are too restrictive and thus make feasible solutions unlikely to be found. Additional arcs should not link any two nodes that represent items which do not share a common machine and they should not link any two nodes that represent the same item. Apparently, disjunctive arcs must not introduce any cycles into the graph structure. In the tests done here, we add disjunctive arcs which link the root nodes of the gozinto–trees and which link nodes with the same successor node in the gozinto–tree. More formally, let $\mathcal{DA}$ denote the set of disjunctive arcs where each element $(h, k) \in \mathcal{DA}$ represents a disjunctive arc pointing from the node with label h to the node with label k. Initially, $\mathcal{DA}$ could be chosen as

$$\mathcal{DA} \stackrel{def}{=} \mathcal{DA}^{ext} \cup \mathcal{DA}^{int} \tag{6.58}$$

where $\mathcal{DA}^{ext}$ is the initial set of disjunctive arcs between root nodes which, by construction, represent external demand and $\mathcal{DA}^{int}$ is the initial set of disjunctive arcs between nodes which represent internal demand. Mathematically, these sets are defined as

$$\mathcal{DA}^{ext} \stackrel{def}{=} \bigcup_{j=1}^{J} \bigcup_{t=1}^{T} \bigcup_{i \in \mathcal{J}_{m_j}} \bigcup_{\tau=t+1}^{T} \{(\omega_1(j,t), \omega_1(i,\tau)) \mid d_{jt} > 0 \wedge d_{i\tau} > 0\}$$

$$\cup \bigcup_{j=1}^{J} \bigcup_{t=1}^{T} \bigcup_{\substack{i \in \mathcal{J}_{m_j} \\ i<j}} \{(\omega_1(j,t), \omega_1(i,t)) \mid d_{jt} > 0 \wedge d_{it} > 0\} \tag{6.59}$$

and

$$\mathcal{DA}^{int} \stackrel{def}{=} \bigcup_{j=1}^{J} \bigcup_{t=1}^{T} \widetilde{\mathcal{DA}}^{int}(j, d_{jt}, \omega_1(j,t)) \tag{6.60}$$

where

$$\widetilde{\mathcal{DA}}^{int}(j, demand, label)$$

$$\stackrel{def}{=} \begin{cases} \emptyset & \text{, if } demand = 0 \\ \bigcup_{i \in \mathcal{P}_j} \bigcup_{\substack{k \in \mathcal{P}_j \cap \mathcal{J}_{m_i} \\ k<i}} \{(\omega_2(j,i,label), \omega_2(j,k,label))\} & \\ \cup \bigcup_{i \in \mathcal{P}_j} \widetilde{\mathcal{DA}}^{int}(i, demand, \omega_2(j,i,label)) & \\ & \text{, otherwise} \end{cases} \tag{6.61}$$

Now, a function σ determines the so-called reference counter of a node which is the number of disjunctive arcs pointing from that node. It can formally be stated as

$$\sigma(label) \stackrel{def}{=} \mid \{(h,k) \in \mathcal{DA} \mid h = label\} \mid \tag{6.62}$$

for $label \in \mathcal{L}$. Note, by construction we have that

$$(h,k) \in \mathcal{DA} \Rightarrow \sigma(h) > \sigma(k) \tag{6.63}$$

holds.

6.6.3 Tabu Search

As pointed out earlier, the basic working principle of the method presented here is to iterate while running a construction and a graph modification phase alternatingly. We are to describe the latter one now where the orientation of one or more disjunctive arcs is changed.

Since it must not happen that redirecting an arc introduces a cycle into the graph Γ^{TS}, in general we cannot choose any of the arcs in the set of disjunctive arcs $\mathcal{DA}$. Due to the definition of $\mathcal{DA}$, the set of valid candidates is

$$\mathcal{VC} \stackrel{def}{=} \{(h,k) \in \mathcal{DA} \mid \sigma(h) - \sigma(k) = 1 \wedge item(h) \neq item(k)\} \tag{6.64}$$

where $item(h)$ gives the item number of the node with label $h \in \mathcal{L}$. Note, the set of disjunctive arcs which could be redirected without introducing a cycle is

$$\mathcal{VC}^{+} \stackrel{def}{=} \{(h,k) \in \mathcal{DA} \mid \sigma(h) - \sigma(k) = 1\} \supseteq \mathcal{VC}. \tag{6.65}$$

But, no arc out of the set $\mathcal{VC}^{+} \backslash \mathcal{VC}$ should ever be redirected, because by construction an arc (h,k) in this set connects two nodes h and k which represent external demand for the same item. By construction again, the demand which is represented by node h occurs earlier than the demand represented by node k and hence it makes no sense to schedule the corresponding production quantities in a different order. Note, there is at most one disjunctive arc originating from each node which is a member of $\mathcal{VC}$, i.e.

$$|\{(h,k) \in \mathcal{VC}\}| \leq 1 \qquad h \in \mathcal{L}. \tag{6.66}$$

Let us assume that $\mathcal{VC} \neq \emptyset$, because otherwise we face an instance for which tabu search plays no role. Nevertheless, the method could be applied to such instances.

Redirecting a disjunctive arc $(h,k) \in \mathcal{VC}$ in a given graph Γ^{TS} gives a new graph $\Gamma^{TS'}$ where the disjunctive arcs are

$$\mathcal{DA}' = (\mathcal{DA} \backslash \{(h,k)\}) \cup \{(k,h)\} \tag{6.67}$$

and the reference counts of the nodes are

$$\sigma(label)' = \begin{cases} \sigma(label) & \text{, if } label \neq k \text{ and } label \neq h \\ \sigma(label)+1 & \text{, if } label = k \\ \sigma(label)-1 & \text{, if } label = h \end{cases} \tag{6.68}$$

for $label \in \mathcal{L}$. Note, the set of valid candidates for choosing the next disjunctive arc to be reversed is

$$\mathcal{VC}' = (\mathcal{VC} \backslash (\{(h,k)\} \cup \mathcal{NB}_{(h,k)})) \cup \{(k,h)\} \cup \overline{\mathcal{NB}}_{(h,k)} \tag{6.69}$$

where

$$\mathcal{NB}_{(h,k)} \stackrel{def}{=} \{(i,j) \in \mathcal{VC} \mid i = k \vee j = h\} \tag{6.70}$$

and

$$\overline{\mathcal{NB}}_{(h,k)} \stackrel{def}{=} \{(i,j) \in \mathcal{DA} \mid \begin{array}{l} (j = k \wedge (i,h) \in \mathcal{NB}_{(h,k)}) \\ \vee (i = h \wedge (k,j) \in \mathcal{NB}_{(h,k)})\} \end{array} \tag{6.71}$$

which can be verified by applying (6.64) to (6.67).

Because we like to perform a tabu search for finding an orientation of the disjunctive arcs that yields a production plan with a low objective function value, a tabu list containing arcs restricts the redirection of arcs. Once an arc $(h,k) \in \mathcal{VC}$ is chosen somehow, we test if (h,k) is contained in the tabu list. If not, the element (k,h) is added to the tabu list which actually is a queue[8] and redirect the arc (h,k). If (h,k) itself is contained in the tabu list, we redirect no arc and append a dummy entry $(0,0) \notin \mathcal{DA}$ to the tabu list instead. Choosing and redirecting arcs can be done $ARCNUM$ times before a new construction phase is started. The parameter $ARCNUM$ is specified by the user. In our tests, we use tabu lists with a static length $LISTLENGTH$ which is a method parameter specified by the user, too. Initially, the tabu list is filled with dummy entries.

What remains to explain is the way to choose a disjunctive arc for reversing its orientation. Perhaps, it would be best to choose every arc $(h,k) \in \mathcal{VC}$ one after the other, change its orientation temporarily, construct a production plan with respect to the new

[8] A queue is a data structure under a first–in first–out regime.

precedence relations, and reinstall the arc's old orientation. After this is done with every valid candidate, we could choose the arc that gave the plan with the lowest objective function value and change its orientation permanently. This is very time–consuming and thus we follow another strategy which differs from what is said above in two important aspects.

First, rather than testing every arc in the set $\mathcal{VC}$ we only test some arcs which are randomly chosen out of $\mathcal{VC}$ with uniform distribution. Let $\widetilde{\mathcal{VC}} \subseteq \mathcal{VC}$ denote the set of arcs that are actually tested where $\mid \widetilde{\mathcal{VC}} \mid \leq ARCSELECT$ holds and the method parameter $ARCSELECT$ is the number of random drawings out of $\mathcal{VC}$ which is specified by the user. Note, some arcs might be chosen more than once and thus $\mid \widetilde{\mathcal{VC}} \mid = ARCSELECT$ is not always true.

Second, rather than changing the orientation of a selected arc $(h, k) \in \widetilde{\mathcal{VC}}$ temporarily and constructing a production plan for this intermediate graph, we make an estimate $\Delta costs_{(h,k)}$ of how the objective function value changes when having the arc (k, h) instead of having (h, k). Eventually, we change the orientation of the arc $(h^*, k^*) \in \widetilde{\mathcal{VC}}$ which is the one with the lowest estimated value, i.e. we choose $(h^*, k^*) \in \widetilde{\mathcal{VC}}$ so that

$$\Delta costs_{(h^*,k^*)} = \min\{\Delta costs_{(h,k)} \mid (h, k) \in \widetilde{\mathcal{VC}}\}. \tag{6.72}$$

Ties are broken arbitrarily.

Note, what is described here is for giving one arc a new orientation. Depending on the user specification, all these lines are to be repeated $ARCNUM$ times giving at most $ARCNUM$ different arcs that are reversed.

Ahead of a formal definition of $\Delta costs_{(h,k)}$, let us discuss the impact of changing the orientation of an arc which gives a motivation for the formula then used.

For $h \in \mathcal{L}$, let $item(h)$ denote the item number of the node with label h again, $demand(h)$ be the size of the demand represented by node h, and $deadline(h)$ a retrieval function that yields the deadline entry at the tuple position 2 in the node h. Changing the orientation of an arc $(h, k) \in \widetilde{\mathcal{VC}}$ affects both, the total holding costs and the total setup costs. Having an arc (k, h) instead of

(h,k) tends to increase the holding costs for $item(k)$ and tends to decrease the holding costs for $item(h)$. An estimate for the impact of the redirection of an arc (h,k) on holding costs thus is:

$$\Delta hc_{(h,k)} \stackrel{def}{=} \begin{cases} demand(k)h_{item(k)} \\ \quad - demand(h)h_{item(h)} \\ \qquad \cdot(1 + deadline(h) - deadline(k)) \\ \qquad\quad , \text{if } deadline(h) \geq deadline(k) \\ \ demand(k)h_{item(k)} \\ \ \cdot(1 + deadline(k) - deadline(h)) \\ -demand(h)h_{item(h)} \\ \qquad , \text{otherwise} \end{cases} \tag{6.73}$$

Whether or not setup costs may change when redirecting an arc (h,k) depends on what other arcs are contained in the set $\mathcal{VC}^+$. If there is, for instance, an arc $(k,i) \in \mathcal{VC}^+$[9] where $item(i) = item(h)$ then the new arc (k,h) helps to save setup costs, because the demand of node h and the demand of node i may be produced in one lot:

$$\Delta sc^{I}_{(h,k)} \stackrel{def}{=} \begin{cases} -s_{item(h)} & , \text{if } \exists (k,i) \in \mathcal{VC}^+ : item(i) = item(h) \\ 0 & , \text{otherwise} \end{cases} \tag{6.74}$$

If there is, for instance, an arc $(k,i) \in \mathcal{VC}^+$ where $item(i) = item(k)$, then the new arc (k,h) may enforce additional setup costs, because the demand of node k and the demand of node i may not be produced in one lot:

$$\Delta sc^{II}_{(h,k)} \stackrel{def}{=} \begin{cases} s_{item(k)} & , \text{if } \exists (k,i) \in \mathcal{VC}^+ : item(i) = item(k) \\ 0 & , \text{otherwise} \end{cases} \tag{6.75}$$

Analogously, there are two other cases:

$$\Delta sc^{III}_{(h,k)} \stackrel{def}{=} \begin{cases} s_{item(h)} & , \text{if } \exists (i,h) \in \mathcal{VC}^+ : item(i) = item(h) \\ 0 & , \text{otherwise} \end{cases} \tag{6.76}$$

[9]Note, this argument only applies when we consider the arc set $\mathcal{VC}^+$ as we do here, but not $\mathcal{DA}$.

$$\Delta sc^{IV}_{(h,k)} \stackrel{def}{=} \begin{cases} -s_{item(k)} & \text{, if } \exists (i,h) \in \mathcal{VC}^{+} : item(i) = item(k) \\ 0 & \text{, otherwise} \end{cases} \tag{6.77}$$

An estimate for the impact of the redirection of an arc (h,k) on setup costs therefore is:

$$\Delta sc_{(h,k)} \stackrel{def}{=} \Delta sc^{I}_{(h,k)} + \Delta sc^{II}_{(h,k)} + \Delta sc^{III}_{(h,k)} + \Delta sc^{IV}_{(h,k)} \tag{6.78}$$

Note, only nodes h and k which represent external demand may give

$$\Delta sc_{(h,k)} \neq 0.$$

In summary, we can now give a definition of $\Delta costs_{(h,k)}$:

$$\Delta costs_{(h,k)} \stackrel{def}{=} \Delta hc_{(h,k)} + \Delta sc_{(h,k)} \tag{6.79}$$

The aspiration criterion used in our tests also makes use of the estimated change of costs. Arcs $(h,k) \in \widetilde{\mathcal{VC}}$ which fulfill

$$\Delta costs_{(h,k)} < 0 \tag{6.80}$$

may be redirected although this operation is set tabu. The motivation for doing so is that redirecting the arc probably gives a production plan with lower objective function value.

6.6.4 Intensification of the Search

Facing a problem for which it is non–trivial to find feasible solutions, it seems to be a good advice to intensify the search in those areas of the solution space which gave feasible solutions with low objective function values. In terms of disjunctive arcs that would mean to keep the orientation of those arcs which lead to improvements or at least to feasible solutions in earlier iterations. Fixing the orientation of some arcs permanently would be a very rigid strategy which cuts off feasible solutions and tends to trap the local search procedure into local optima. Thus, we follow a less restrictive approach. Consider

$$\mathcal{DA}^{*} \stackrel{def}{=} \{(h,k) \mid (h,k) \in \mathcal{DA} \vee (k,h) \in \mathcal{DA}\} \tag{6.81}$$

which is the set of all disjunctive arcs that might appear during run–time. Note, this set does not change during run–time and can therefore be computed once and for all using the initial set of disjunctive arcs $\mathcal{DA}$ as a basis. Attached to each arc $(h,k) \in \mathcal{DA}^*$ we maintain a positive arc weight $weight_{(h,k)}$. Its interpretation is that the higher the weight of an arc (h,k) the more promising it is to have a disjunctive arc pointing from node h to node k. These weights are now used to tune the $\Delta costs_{(h,k)}$–values which are defined in (6.79):

$$\Delta costs'_{(h,k)} = \begin{cases} \Delta costs_{(h,k)} \frac{weight_{(h,k)}}{weight_{(k,h)}} & \text{, if } \Delta costs_{(h,k)} > 0 \\ \Delta costs_{(h,k)} \frac{weight_{(k,h)}}{weight_{(h,k)}} & \text{, otherwise} \end{cases} \tag{6.82}$$

The effect of this modification is the following. Let $(h,k) \in \widetilde{\mathcal{VC}}$. Suppose that $\Delta costs_{(h,k)}$ is positive which indicates that reversing the arc (h,k) increases the objective function value. If $weight_{(h,k)} > weight_{(k,h)}$ then past performance indicates that having the arc (h,k) is more promising than having the arc (k,h). For $\Delta costs'_{(h,k)} > \Delta costs_{(h,k)} > 0$ the modified value signals that there is no clue that selecting the arc (h,k) out of $\widetilde{\mathcal{VC}}$ and reversing it is a good choice. If $weight_{(h,k)} < weight_{(k,h)}$ then the run–time history tells us that it seems to be better having the arc (k,h) instead of (h,k). Due to $\Delta costs_{(h,k)} > \Delta costs'_{(h,k)} > 0$ the chance for selecting and redirecting the arc (h,k) increases. Suppose now that $\Delta costs_{(h,k)} \leq 0$ holds which means that redirecting the arc (h,k) probably is an improvement step. If $weight_{(k,h)} > weight_{(h,k)}$, past performance underscores this point. And thus, $\Delta costs'_{(h,k)} < \Delta costs_{(h,k)} < 0$ puts some more emphasis on the fact that redirecting the arc (h,k) seems to be a good idea. If $weight_{(h,k)} > weight_{(k,h)}$, we get $\Delta costs_{(h,k)} < \Delta costs'_{(h,k)} < 0$ which reflects that the objective function value might decrease when reversing the arc (h,k), but former iterations indicate that the arc (h,k) should be preferred to (k,h).

All that is left to discuss now is how to choose and update the arc weights during run–time. Initially, we set $weight_{(h,k)} = 1$ for $(h,k) \in \mathcal{DA}^*$. For the first iteration these values are of no relevance, because the first time the construction scheme runs, the initial graph structure Γ^{TS} is used. After the very first production plan is constructed we use the initial weights to select at most $ARCNUM$ disjunctive arcs, reverse them, and start the construction scheme for the second time. Once the production plan number $n \geq 2$ is generated, we update the arc weights before we modify the current graph Γ^{TS} and begin the next construction phase. Updating the arc weights is done as follows. Remember that we have reversed (at most) $ARCNUM$ arcs at the beginning of the iteration number n. Let the set of arcs with new orientation be denoted as $\mathcal{NA}_n$ where $(h,k) \in \mathcal{NA}_n$ means that the disjunctive arc (h,k) is in the graph used for constructing production plan number n and the arc (k,h) is in the graph used for constructing the plan in the iteration $n-1$. If the construction phase number n finds no feasible plan nothing happens. If the construction phase gives a feasible plan which does not improve the current best solution, we update

$$weight_{(h,k)} = weight_{(h,k)} + 1 \tag{6.83}$$

for $(h,k) \in \mathcal{NA}_n$. If the construction phase number n results in a feasible plan that improves the current best solution, we compute

$$weight_{(h,k)} = weight_{(h,k)} + n \tag{6.84}$$

for $(h,k) \in \mathcal{NA}_n$. So, whenever redirecting arcs leads to a feasible solution the weight of the most recent arcs is increased. The later an improvement is reached, the more promising it seems to keep the orientation of those arcs which are responsible for the improvement.

6.6.5 Modifications of the Construction Scheme

The underlying logic of the construction scheme that generates a production plan where all precedence relations in a given data structure Γ^{TS} are respected is very much alike the principles described in Section 6.2. While moving from machine to machine and backwards from period to period again, setup states and production quantities are computed. What differs is the way production quantities are determined, because this is where the precedence relations defined by the disjunctive arcs comes in.

To give a formal specification of how the construction scheme works, let us introduce some notation. For each node with label $h \in \mathcal{L}$, be *item*(h), *deadline*(h), and *demand*(h) as defined above. Furthermore, let *preds*(h) denote the set of labels of immediately preceding nodes in the gozinto–tree. The tuples representing these nodes can be found at the last position in the tuple which represents the node h. Additionally, let us have a counter $counter_h$ for each node which simply counts the number of nodes which are connected with node h via a disjunctive arc pointing from node h and which are already scheduled. Initially, $counter_h = 0$ for $h \in \mathcal{L}$. Disregarding the disjunctive arcs for a moment, $\mathcal{NL}$ is the set of all node labels among which the next node to schedule is to be found. Initially, this is the set of all root node labels of the gozinto–trees, i.e.

$$\mathcal{NL} \stackrel{def}{=} \{\omega_1(j,t) \mid j \in \{1,\ldots,J\} \wedge t \in \{1,\ldots,T\} \wedge d_{jt} > 0\}. \tag{6.85}$$

The deadline entries for nodes which represent internal demand need to be updated when the unique successor node in the gozinto–tree is scheduled. Let $h \otimes_2 t$ denote the operation which temporarily updates the deadline entry of the node with label h (which is the entry at position two in the tuple that represents node h) with the value t. The old values (which can be found at position three in the tuple) are restored after a construction phase terminates and before the next tabu search phase starts. Similar to the *construct*$(t, \Delta t, m)$–procedure given in Section 6.2 we are now able to specify a *DA–construct*$(t, \Delta t, m)$–method which additionally takes the precedence relations given by the disjunctive arcs into account. All variables have the same initial values as before. Additionally, we have variables Q_{jt} for $j = 1, \ldots, J$ and $t = 1, \ldots, T$ which are initialized with zero. For the sake of convenience, let us assume $Q_{j(T+1)} = 0$ for $j = 1, \ldots, J$. These variables will give the upper bound for the production quantities of item j in period t that is due to the precedence relations defined by the disjunctive arcs. Again, we do not allow lot splitting. Calling *DA–construct*$(T, 1, 1)$ initiates the construction of a production plan. See Table 6.31 for the details.

$\mathcal{DS}_{mT} := \{k \in \mathcal{NL} \mid m_{item(k)} = m \wedge \sigma(k) - counter_k = 0\}$.
if ($\mathcal{DS}_{mT} = \emptyset$)
    $j_{mT} := 0$.
else
    choose $h \in \mathcal{DS}_{mT}$.
    $j_{mT} := item(h)$.
    $y_{j_{mT}T} := 1$.
if ($m = M$)
    $DA\text{-}construct(T, 0, 1)$.
else
    $DA\text{-}construct(T, 1, m + 1)$.

Table 6.31: Evaluating $DA\text{-}construct(T, 1, \cdot)$

When comparing Table 6.31 with Table 6.1 one can see that the decision set $\mathcal{DS}_{mT} \subseteq \mathcal{NL}$ is new. It contains the labels of those nodes which might be scheduled next. Choosing one of these labels gives the item machine m is set up for at the end of period T. In other words, choosing a label $h \in \mathcal{DS}_{mT}$ replaces the choice $j_{mT} \in \mathcal{I}_{mT}$. The way labels are chosen out of $\mathcal{DS}_{mT}$ is of minor importance for the understanding and will thus be discussed after the *DA-construct*-procedure is totally outlined. How to evaluate calls of the form $DA\text{-}construct(t, 0, \cdot)$ is described in Table 6.32.

The idea behind this piece of code is to meet all demand for the item j_{mt} that is represented by nodes which can be scheduled without violating the precedence relations. This quantity is denoted as $Q_{j_{mt}t}$. Table 6.33 describes how to determine this value. The rest of the code closely relates to what is defined in Section 6.2. All that is new is that $Q_{j_{mt}t}$ gives an upper bound for the production quantities. Eventually, when $DA\text{-}construct(t, 1, \cdot)$ is called, we need Table 6.34.

The choice of j_{mt} is done so that lots are not split. If lot splitting should be allowed, testing $j_{m(t+1)} \neq 0$ and $Q_{j_{m(t+1)}(t+1)} > 0$ should be left out. From then on we pass the code that computes the production quantities in period $t + 1$ which is fairly the same as

compute $Q_{j_{mt}t}$.
for $j \in \mathcal{J}_m$
$CD_{jt} := \min\left\{CD_{j(t+1)} + \tilde{d}_{jt}, \max\{0, nr_j - \sum_{\tau=t+1}^{T} q_{j\tau}\}\right\}$.
$Q_{jt} := \min\left\{Q_{jt} + Q_{j(t+1)}, \max\{0, nr_j - \sum_{\tau=t+1}^{T} q_{j\tau}\}\right\}$.
if $(j_{mt} \neq 0)$
$q_{j_{mt}t} := \min\left\{CD_{j_{mt}t}, Q_{j_{mt}t}, \frac{RC_{mt}}{p_{j_{mt}}}\right\}$.
$Q_{j_{mt}t} := Q_{j_{mt}t} - q_{j_{mt}t}$.
$CD_{j_{mt}t} := CD_{j_{mt}t} - q_{j_{mt}t}$.
$RC_{mt} := RC_{mt} - p_{j_{mt}}q_{j_{mt}t}$.
for $i \in \mathcal{P}_{j_{mt}}$
if $(t - v_i > 0$ and $q_{j_{mt}t} > 0)$
$\tilde{d}_{i(t-v_i)} := \tilde{d}_{i(t-v_i)} + a_{ij_{mt}}q_{j_{mt}t}$.
if $(m = M)$
DA-construct$(t-1, 1, 1)$.
else
DA-construct$(t, 0, m+1)$.

Table 6.32: Evaluating *DA-construct*$(t, 0, \cdot)$ where $1 \leq t \leq T$

what is given in Table 6.3. Again, the consideration of $Q_{j_{mt}(t+1)}$ is new.

Evaluating *DA-construct*$(0, 1, \cdot)$ follows almost the same lines as *DA-construct*$(t, 1, \cdot)$ where $t > 0$. All that changes is that the initial setup state j_{m0} is already known and hence we have no choice for it. A call to *DA-construct*$(0, 0, \cdot)$ terminates the construction scheme and tests for feasibility as described in Section 6.2.

The construction scheme which takes the disjunctive arcs into account is now close to be completely defined. What is missing is a rule that tells us how to choose a node label h out of $\mathcal{DS}_{mt}$ in Tables 6.31 and 6.34. Attached to each node $h \in \mathcal{DS}_{mt}$ is a priority value defined as

$$priority_{ht} \stackrel{def}{=} demand(h)h_{item(h)}(1 + deadline(h) - t) \quad (6.86)$$

---

while $(\exists h \in \mathcal{DS}_{mt} : item(h) = j_{mt} \wedge deadline(h) \geq t + \Delta t)$
    $Q_{j_{mt}(t+\Delta t)} := Q_{j_{mt}(t+\Delta t)} + demand(h).$
    $\mathcal{NL} := (\mathcal{NL} \backslash \{h\}) \cup preds(h).$
    for $k \in preds(h)$
        $k \otimes_2 (t + \Delta t - v_{item(k)}).$
    for $k \in \mathcal{L}$ where $(k, h) \in \mathcal{DA}$
        $counter_k := counter_k + 1.$
    $\mathcal{DS}_{mt} := \{k \in \mathcal{NL} \mid m_{item(k)} = m \wedge \sigma(k) - counter_k = 0\}.$

Table 6.33: Computing $Q_{j_{mt}(t+\Delta t)}$

which measures the holding costs that are charged if the demand represented by node h is not chosen to be met. Note, nodes with a deadline prior to period t do not have a positive value. And, the longer we delay meeting the demand represented by a node the higher is the priority value of this particular node. To limit the computational effort to find a node with a high priority value we do not determine the priority rule for every node in $\mathcal{DS}_{mt}$. Instead we draw $NODESELECT > 0$ labels out of $\mathcal{DS}_{mt}$ with uniform distribution (some labels might be drawn more than once). Among these the node with the highest priority value is chosen. The integral method parameter $NODESELECT$ is specified by the user.

6.6.6 An Example

Once again, let us consider the example given in Subsection 6.3.5. Furthermore, assume the data structure depicted in Figure 6.5. The upper indices of the item numbers are the labels as computed using formulae (6.56) and (6.55), respectively. For instance, the node that represents the external demand for item 4 in period 8 has label 5.

$\mathcal{DS}_{mt} := \{k \in \mathcal{NL} \mid m_{item(k)} = m \wedge \sigma(k) - counter_k = 0\}$.
if $(\mathcal{DS}_{mt} = \emptyset$ or $(j_{m(t+1)} \neq 0$ and $Q_{j_{m(t+1)}(t+1)} > 0))$
    $j_{mt} := j_{m(t+1)}$.
else
    choose $h \in \mathcal{DS}_{mt}$.
    $j_{mt} := item(h)$.
compute $Q_{j_{mt}(t+1)}$.
if $(j_{mt} \neq 0)$
    $y_{j_{mt}t} := 1$.
    if $(j_{mt} \neq j_{m(t+1)})$
        $q_{j_{mt}(t+1)} := \min\left\{CD_{j_{mt}(t+1)}, Q_{j_{mt}(t+1)}, \frac{RC_{m(t+1)}}{p_{j_{mt}}}\right\}$.
        $Q_{j_{mt}(t+1)} := Q_{j_{mt}(t+1)} - q_{j_{mt}(t+1)}$.
        $CD_{j_{mt}(t+1)} := CD_{j_{mt}(t+1)} - q_{j_{mt}(t+1)}$.
        $RC_{m(t+1)} := RC_{m(t+1)} - p_{j_{mt}} q_{j_{mt}(t+1)}$.
        for $i \in \mathcal{P}_{j_{mt}}$
            if $(t + 1 - v_i > 0$ and $q_{j_{mt}(t+1)} > 0)$
                $\tilde{d}_{i(t+1-v_i)} := \tilde{d}_{i(t+1-v_i)} + a_{ij_{mt}} q_{j_{mt}(t+1)}$.
if $(m = M)$
    $DA\text{-}construct(t, 0, 1)$.
else
    $DA\text{-}construct(t, 1, m + 1)$.

Table 6.34: Evaluating $DA\text{-}construct(t, 1, \cdot)$ where $1 \leq t < T$

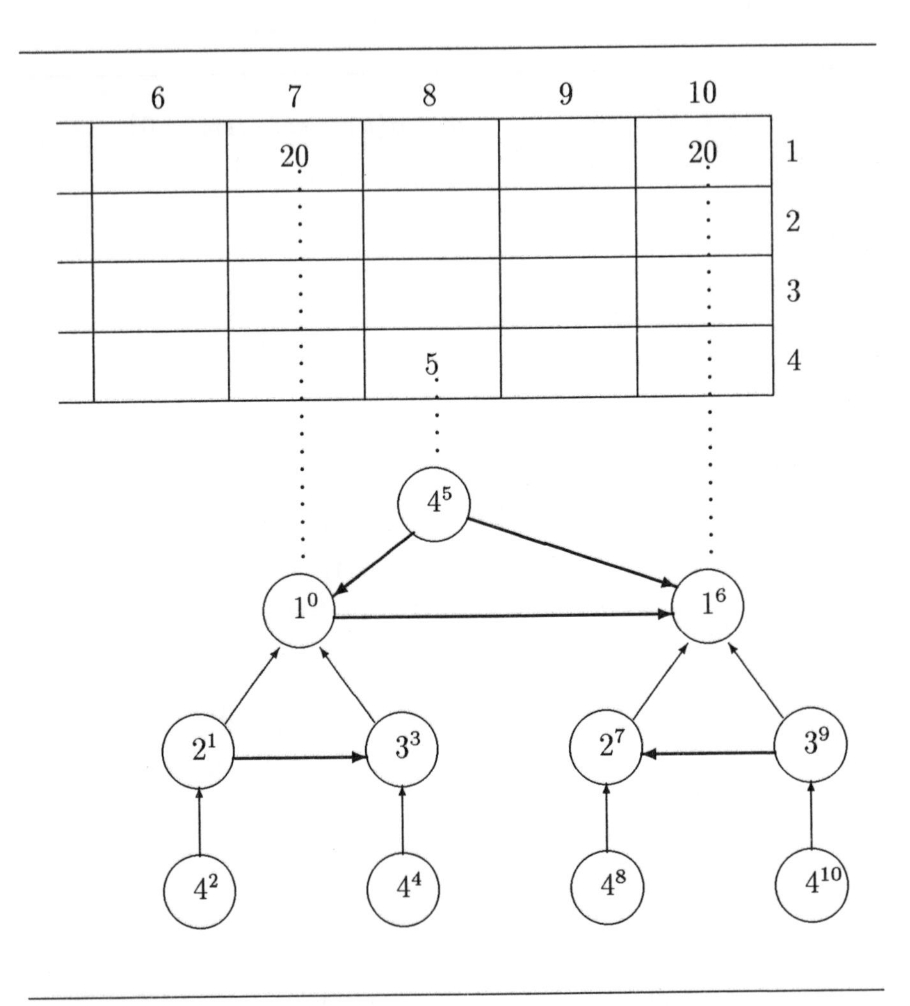

Figure 6.5: The Data Structure for the Example

More formally, the data structure can be given as

$$
\begin{aligned}
\Gamma^{TS} = (\{(1,7,7,20,0, \quad & \{(2,6,6,20,1,\{(4,5,5,20,2,\emptyset)\}),\\
& (3,6,6,20,3,\{(4,5,5,20,4,\emptyset)\})\})\\
(4,8,8,5,5,\emptyset),&\\
(1,10,10,20,6,\{ & (2,9,9,20,7,\{(4,8,8,20,8,\emptyset)\}),\\
& (3,9,9,20,9,\{(4,8,8,20,10,\emptyset)\})\})\\
\},\mathcal{DA})&
\end{aligned}
$$

where $\mathcal{DA} = \{(0,6),(5,0),(5,6),(1,3),(9,7)\}$ is the set of disjunctive arcs and thus $\sigma(h)$ for $h \in \{0,\ldots,10\}$ is as given in Table 6.35. For instance, the node which represents the external demand for item 1 in period 7 is a tuple with label 0. We have $item(0) = 1$, $deadline(0) = 7$, $demand(0) = 20$, and $preds(0) = \{1,3\}$. Since $nodes_1 = 5$ the node with label 0 is the root node of a gozinto–tree with five nodes.

$h =$	0	1	2	3	4	5	6	7	8	9	10
$\sigma(h) =$	1	1	0	0	0	2	0	0	0	1	0

Table 6.35: The $\sigma(h)$–Values of the Example

Note, $\mathcal{VC} = \{(5,0),(1,3),(9,7)\}$ is the set of disjunctive arcs that are allowed to be redirected (see (6.64)). The arc $(5,6)$ is not included in this set, because changing the orientation of this arc would introduce a cycle. Redirecting the arc $(0,6)$ does not make sense, since nodes 0 and 6 represent external demand for the same item and $deadline(0) < deadline(6)$.

Table 6.36 gives a protocol of the first few steps of the construction scheme. A possible outcome is given in Figure 6.2.

Step	$(t, \Delta t, m)$	$\vec{\tilde{d}}_{jt}$ $\vec{CD}_{j(t+\Delta t)}$	$\mathcal{NL}$	$\mathcal{DS}_{mt}$	j_{mt}	$\vec{Q}_{j(t+\Delta t)}$	$q_{j_{mt}(t+\Delta t)}$
1	(10,1,1)	(20,0,0,0) (0,0,0,0)	$\{0, 5, 6\}$	$\{6\}$	1	(0,0,0,0)	
2	(10,1,2)	(20,0,0,0) (0,0,0,0)	$\{0, 5, 6\}$	$\emptyset$	0	(0,0,0,0)	
3	(10,0,1)	(20,0,0,0) (20,0,0,0)	$\{0, 5, 6\}$	$\{6\}$		(20,0,0,0)	$q_{1T} = 15$
4	(10,0,2)	(20,0,0,0) (5,0,0,0)	$\{0, 5, 7, 9\}$	$\{7\}$		(5,0,0,0)	
5	(9,1,1)	(0,15,15,0) (5,0,0,0)	$\{0, 5, 7, 9\}$	$\{0\}$	1	(5,0,0,0)	
6	(9,1,2)	(0,15,15,0) (5,0,0,0)	$\{0, 5, 7, 9\}$	$\{7\}$	2	(5,0,0,0)	$q_{2T} = 0$
7	(9,0,1)	(0,15,15,0) (5,0,0,0)	$\{0, 5, 7, 9\}$	$\{0\}$		(5,0,0,0)	$q_{19} = 5$
8	(9,0,2)	(0,15,15,0) (0,15,15,0)	$\{0, 5, 7, 9\}$	$\{7\}$		(0,20,0,0)	$q_{29} = 15$
9	(8,1,1)	(0,5,5,20) (0,0,15,0)	$\{0, 5, 8, 9\}$	$\{0, 8\}$	1	(0,5,0,0)	
10	(8,1,2)	(0,5,5,20) (0,0,15,0)	$\{0, 5, 8, 9\}$	$\{9\}$	2	(0,5,0,0)	
11	(8,0,1)	(0,5,5,20) (0,0,15,20)	$\{0, 5, 8, 9\}$	$\{0, 8\}$		(0,5,0,0)	$q_{18} = 0$
12	(8,0,2)	(0,5,5,20) (0,5,20,20)	$\{0, 5, 8, 9\}$	$\{9\}$		(0,5,0,0)	$q_{28} = 5$
13	(7,1,1)	(20,0,0,5) (0,0,20,20)	$\{0, 5, 8, 9\}$	$\{0, 8\}$	1	(0,0,0,0)	
14	(7,1,2)	(20,0,0,5) (0,0,20,20)	$\{0, 5, 8, 9\}$	$\{9\}$	3	(0,0,20,0)	$q_{38} = 10$
...							

Table 6.36: A Protocol of the Construction Scheme of the Disjunctive Arc Based Procedure

Some additional remarks will stress a couple of interesting points.

Step 1: Initially, $\mathcal{NL} = \{0, 5, 6\}$ which is the set of all root node labels. However, only node 6 is a valid choice which is due to the precedence relations among the root nodes.

Step 2: $\mathcal{DS}_{2T} = \emptyset$, because all nodes in $\mathcal{NL}$ represent items to be produced on machine 1.

Step 3: The node with label 6 is eliminated from the set $\mathcal{NL}$. Hence, $counter_0 = 1$ and $counter_5 = 1$. Additionally, $demand(6) =$ 20 units of item 1 become available for being scheduled. Thus, $Q_{1T} = 20$. Furthermore, the immediately preceding nodes of node 6 in the gozinto–tree, namely nodes 7 and 9, are included into the set $\mathcal{NL}$ (see Step 4).

Step 8: Node 7 is selected. Hence, 20 units for item 2 become available for being scheduled. Note, although there is demand for item 3 as well ($CD_{39} = 15$), we have $Q_{39} = 0$.

Step 9: Both, node 0 and node 8 could be chosen next. We decide to select node 0 which means that $j_{18} = 1$. We have $deadline(5)$ $= 8$, but node 5 is not contained in the set $\mathcal{DS}_{18}$. This is because $\sigma(5) - counter_5 = 2 - 1 = 1 \neq 0$. Node 5 is available only if node 0 is scheduled before.

Step 10: Despite $\mathcal{DS}_{28} = \{9\}$ and $item(9) = 3$, we choose $j_{28} = 2$ to avoid a splitting of the lot for item 2 ($Q_{29} > 0$).

Step 11: Assume that we would have chosen node 8 instead of node 0 in Step 9 where $item(8) = 4$. Then, Q_{48} would equal $CD_{48} = 20$ which is the sum of internal and external demand for item 4 in period 8, although node 5 which is the node representing the external demand for item 4 in period 8 was not scheduled. As a consequence, producing the lot for item 4 would last on until all demand is met. Suppose we would have, say $demand(8) = 15$, which would be the case if the external demand for item 1 in period $T = 10$ would be $d_{1T} = 15$. Then, $Q_{48} = 15 < 20 = CD_{48}$ and only the internal demand for item 4 in period 8 should be met, but not the external demand. This points out the relevance for having Q_{jt} as an upper bound for the production quantity of item j in period t.

Step 14: Now, we have no other choice than setting machine 2 up for item 3.

6.6.7 Experimental Evaluation

To test the disjunctive arc based tabu search we execute 1,000 iterations to allow a fair comparison with other methods and choose the following method parameters: $ARCNUM = 1$, $LISTLENGTH = 5$, $ARCSELECT = 20$, and $NODESELECT = 3$. Tables B.10 to B.12 in the Appendix B provide detailed results in the same format that is used in preceding sections, i.e. we find the average deviation from the optimum objective function value, the worst case deviation, and the number of instances for which a feasible solution is found.

Again, we aggregate the data to get a feeling of which parameter levels give good and which give bad results. To start with, Table 6.37 deals with changes of the number of machines. We can see that increasing the number of machines comes along with an increase of the average deviation. Also, the average run–time increases significantly, but the infeasibility ratio declines.

	$M = 1$	$M = 2$
Average Deviation	16.98	18.18
Infeasibility Ratio	36.50	34.75
Average Run–Time	0.41	0.58

Table 6.37: The Impact of the Number of Machines on the Performance

For studying the influence of the gozinto–structure complexity, let us have a look at Table 6.38. As we see, a higher complexity reduces the average deviation from the optimum objective function value. Unfortunately, it also raises the infeasibility ratio and the run–time effort drastically.

Different demand patterns are concerned about in Table 6.39. All three performance measures are heavily affected by the pattern

	$\mathcal{C} = 0.2$	$\mathcal{C} = 0.8$
Average Deviation	18.87	16.02
Infeasibility Ratio	30.10	41.34
Average Run–Time	0.45	0.56

Table 6.38: The Impact of the Gozinto–Structure Complexity on the Performance

of demand. While sparsely–filled demand matrices result in reasonable performance, the outcome for instances with many non–zeroes in the demand matrix is disappointing.

	$(T_{macro}, T_{micro}, T_{idle}) =$		
	$(10, 1, 5)$	$(5, 2, 2)$	$(1, 10, 0)$
Average Deviation	34.65	18.05	7.17
Infeasibility Ratio	52.42	33.64	20.95
Average Run–Time	0.69	0.52	0.37

Table 6.39: The Impact of the Demand Pattern on the Performance

Table 6.40 reveals whether or not the cost structure has an impact on the performance. The best average deviation from the optimum is gained when setup costs are not too low and not too high. Extreme cost ratios give decidedly worse results. For a small ratio of setup and holding costs the method yields the highest infeasibility ratio of all cost structures. The run–time performance is unaffected by the cost structure.

Table 6.41 investigates the impact of the capacity utilization. Remarkable to note, a high utilization gives quite a low average deviation from the optimum whereas the disjunctive arc based tabu search method fails for low capacity usage. The infeasibility ratio is positively correlated with the capacity utilization. The run–time appears to be almost unaffected by different parameter levels.

In summary, 368 out of the 1,033 instances in the test–bed re-

	$COSTRATIO =$		
	5	150	900
Average Deviation	17.52	12.90	22.36
Infeasibility Ratio	39.13	33.91	33.82
Average Run–Time	0.49	0.50	0.50

Table 6.40: The Impact of the Cost Structure on the Performance

	$U = 30$	$U = 50$	$U = 70$
Average Deviation	21.73	18.30	10.72
Infeasibility Ratio	28.09	35.47	44.20
Average Run–Time	0.50	0.51	0.48

Table 6.41: The Impact of the Capacity Utilization on the Performance

main unsolved. The overall infeasibility ratio thus is 35.62% which is really bad. The average run–time is 0.50 CPU–seconds. The overall average deviation from the optimum objective function value is 17.59%.

As a result, it seems that introducing disjunctive arcs as we do imposes too many restrictions. Instances can hardly be solved as it is indicated by the high infeasibility ratio. Thus, we test what happens if we introduce no disjunctive arcs at all. This switches off the tabu search logic and gives a sampling procedure to which we like to refer to as the simple sampling method. Table 6.42 provides the overall average results for running the simple sampling method with all method parameters kept as they are.

Running the simple sampling method clearly improves all performance measures. A detailed analysis, however, is not performed, because the simple sampling method is closely related to the regret based sampling method discussed in Section 6.3.

	Average Deviation	Infeasibility Ratio	Average Run–Time
Disjunctive Arc Based Tabu Search	17.59	35.62	0.50
Simple Sampling	9.20	17.62	0.47

Table 6.42: The Impact of the Disjunctive Arcs on the Performance

6.7 Demand Shuffle

In this section we get acquainted with another novel heuristic to attack the PLSP–MM. It is a random sampling method again that is combined with a data structure which affects the construction of production plans. Due to some problem specific data structure manipulations each iteration has a tendency to yield a different production plan. Hence, the approach can be seen as a diversified search for production plans with low objective function value.

Results of preliminary tests for the multi–level, single–machine case are given in [Kim94c].

6.7.1 Basic Ideas

Upper bounds Q_{jt} for the sizes of lots for items $j = 1, \ldots, J$ in periods $t = 1, \ldots, T$ bring in a means to control the generation of production plans.[10] Roughly speaking, the three ingredients of the method proposed here are a rule to compute the Q_{jt}–values, a construction scheme respecting these upper bounds, and a way to enforce that the upper bounds will likely change from iteration to iteration.

[10] Compare Section 6.6 where additional precedence constraints among the items lead to upper bounds for the production quantities.

6.7.2 The Data Structure

A graphical representation derived from the gozinto–trees helps to compute upper bounds Q_{jt} for the production quantities. The fundamental idea is that for each positive entry in the external demand matrix a gozinto–tree–like structure not only contains the information what (external or internal) demand occurs for what item, but also a deadline at which this demand is to be met. An important difference to the disjunctive arc based method presented in Section 6.6 will be that these deadline entries will not be updated during the construction phase,[11] but will be determined before the construction of a production plan starts.

More formally, the initial data structure Γ^{DS} can be defined as[12]

$$\Gamma^{DS} \stackrel{def}{=} \bigcup_{j=1}^{J} \bigcup_{t=1}^{T} \tilde{\Gamma}_j^{DS}(t, d_{jt}, \omega_1(j,t), 0, -1) \tag{6.87}$$

where the function ω_1 is defined by (6.56) and computes unique labels for the root nodes of the gozinto–trees. The function $\tilde{\Gamma}_j^{DS}$ is similar to the function $\tilde{\Gamma}_j^{TS}$ given in Subsection 6.6.2. That is,

$$\tilde{\Gamma}_j^{DS}(period, demand, label, path, successor)$$

$$\stackrel{def}{=} \begin{cases} \emptyset & \text{, if } demand = 0 \\ \{(\; j, period, period, demand, label, \\ \quad \alpha(j, path), \\ \quad successor, \\ \quad \bigcup_{i \in \mathcal{P}_j} \tilde{\Gamma}_i^{DS}(\; period - v_i, \\ \qquad a_{ij} demand, \\ \qquad \omega_2(j, i, label), \\ \qquad \alpha(j, path), \\ \qquad label)\} & \text{, otherwise} \end{cases} \tag{6.88}$$

[11] As we will see, there is one exception where deadline entries are updated during the construction phase.

[12] The upper index DS stands for Demand Shuffle.

where the function ω_2 is defined by (6.55) to compute unique labels for the internal demand nodes. Again, each node is represented by a tuple. For retrieving the initial deadline entry at position three of a node h, $deadline^{L4L}(h)$ will be used. What is new are the entries at positions six and seven in the tuple. At the sixth position we find an encoding of the path from the root node to the current node in the gozinto–tree. This encoding scheme is based on a J–ary code where

$$\alpha(j, path) \stackrel{def}{=} path \cdot (J+1) + j. \tag{6.89}$$

Note, the function α assigns a unique number to every node in a gozinto–tree. In the sequel, let us use $path(h)$ to access this value when the node has label $h \in \mathcal{L}$ where $\mathcal{L}$ equals the interval given in (6.57). Also note, in the data structure Γ^{DS} we may have several nodes with the same path value. This happens to be if there is more than one period in which external demand occurs for the same item. At position seven in the tuple we have the label of the immediate successor in the gozinto–tree. The value -1 marks root nodes. For $h \in \mathcal{L}$, let $succ(h)$ be a retrieval function for that value. The explanation of all the other information contained in a node equals the one given in Subsection 6.6.2 and should therefore be omitted here.

6.7.3 Data Structure Manipulations

An upper bound for the production quantity of an item j in a period t is

$$Q_{jt} \stackrel{def}{=} \sum_{\substack{h \in \mathcal{L} \\ item(h)=j \\ deadline(h) \geq t}} demand(h) - \sum_{\tau=t+1}^{T} q_{j\tau}. \tag{6.90}$$

Since the deadline entries in the initial data structure are computed under a lot–for–lot assumption, the cumulative demand for an item always is less than or equal to this upper bound in the first run of the construction scheme. But, modifying the deadline values after each construction phase (e.g. by assigning an earlier deadline) may give upper bounds smaller than the cumulative demand and therefore affects the construction of production plans.

What we need to discuss now is the way the deadline entries are modified. Choosing some nodes at pure random and updating their deadline values with random numbers out of the interval $[1, T]$ would be an easy–to–implement possibility, though not a very insightful one. The strategy that we use can be outlined as follows: First, choose a node on the basis of some priority rule. Then, decide if the deadline entry should be increased or decreased using problem specific insight. And finally, compute the new deadline. For the sake of convenience, let us from now on in this subsection assume that initial inventories do not exist.[13] For all nodes in the data structure the deadline entries can thus be assumed to be positive.

The process of modifying deadlines becomes a bit more illustrative if we imagine a table with J rows and T columns similar to a demand matrix. At a position (j, t) of that matrix we find the sum of all demands that are represented by nodes with label h for which $item(h) = j$ and $deadline(h) = t$ holds. Modifying the deadline entry of a node then corresponds to shifting some demand to the left or to the right, respectively, to give a new table. When doing so over and over again, it looks like shuffling the demand. This coins the name demand shuffle as we like to call the heuristic as a whole.

Before more details are given, let us discuss what side constraints the modification of deadlines should respect. As preliminary studies revealed, shifting demands should not be done arbitrarily. Since we face scarce capacities, the condition

$$\sum_{\tau=1}^{t} C_{m\tau} \geq \sum_{\substack{h \in \mathcal{L} \\ m_{item(h)}=m \\ deadline(h) \leq t}} p_{item(h)} demand(h) \tag{6.91}$$

should be guaranteed to be true for $m = 1, \ldots, M$ and $t = 1, \ldots, T$. Initially, this ought to be true, because no feasible solution exists, otherwise. Note, we only need to test this condition when some demand is shifted to the left. Right shift operations cannot lead to a violation of this restriction.

[13]We return to that point in Section 10.3 again.

Furthermore, the gozinto–structure defines some precedence relations that should be taken into account. Shifting some demand to the left (which corresponds to decreasing the deadline entry of a node) should respect the internal demand that is to be met. Stating this mathematically, a lower bound for a valid deadline entry of a node $h \in \mathcal{L}$ is:

$$deadline^{I}_{LB}(h)$$

$$\stackrel{def}{=} \begin{cases} \max\{\, deadline(k) + v_{item(k)} \\ \qquad \mid k \in preds(h)\} & \text{, if } preds(h) \neq \emptyset \\ 1 & \text{, otherwise} \end{cases} \tag{6.92}$$

Something similar applies when shifting demand to the right (which corresponds to increasing the deadline entry of a node). This time we compute an upper bound for the deadline entry of a node $h \in \mathcal{L}$ by

$$deadline^{I}_{UB}(h)$$

$$\stackrel{def}{=} \begin{cases} deadline(succ(h)) - v_{item(h)} & \text{, if } succ(h) \neq -1 \\ deadline^{L4L}(h) & \text{, otherwise} \end{cases} \tag{6.93}$$

Note, initially

$$deadline^{I}_{LB} = deadline(h) = deadline^{I}_{UB}(h)$$

holds for $h \in \{k \in \mathcal{L} \mid preds(k) \neq \emptyset\}$ and

$$deadline^{I}_{LB} = 1 \leq deadline(h) = deadline^{I}_{UB}(h)$$

is valid for $h \in \{k \in \mathcal{L} \mid preds(k) = \emptyset\}$.

In addition, we utilize the fact that without loss of generality two external demands for the same item are met so that production fulfilling the early demand takes place before the production for the late demand does (this is often called the earliest due date rule). In a multi–level case this result can be extended for internal demands in the following way: If two internal demands have the same position in a gozinto–tree representation then the earliest due date rule applies for these, too. In summary and in terms of our representation of a PLSP–instance, we can now state that only those production plans need to be considered where for each node with label $h \in \mathcal{L}$ its demand is met by production in a certain period only when the demand for all nodes in the set

$$\mathcal{LW}(h) \stackrel{def}{=} \{k \in \mathcal{L} \mid \begin{array}{l} path(k) = path(h) \wedge \\ deadline^{L4L}(k) < deadline^{L4L}(h)\} \end{array} \tag{6.94}$$

is met by production no later than that period. For $h \in \mathcal{L}$ we easily verify the property

$$k_1, k_2 \in \mathcal{LW}(h) \text{ and } k_1 < k_2 \Rightarrow \mathcal{LW}(k_1) \subset \mathcal{LW}(k_2) \subset \mathcal{LW}(h) \tag{6.95}$$

which enables us to avoid testing the complete set $\mathcal{LW}(h)$ to see if the earliest due date rule is respected when shifting the demand represented by node h to the left. It suffices to check the node with label $leftwing(h)$ where

$$leftwing(h) \stackrel{def}{=} \begin{cases} \max(\mathcal{LW}(h)) & \text{, if } \mathcal{LW}(h) \neq \emptyset \\ -1 & \text{, otherwise} \end{cases} \tag{6.96}$$

and where the value -1 indicates that no such test is to be made, because there simply is no demand that should be met prior to the demand represented by node h according to the earliest due date rule. As a result we have another lower bound for the deadline entry of a node $h \in \mathcal{L}$:

$$deadline_{LB}^{II}(h) \stackrel{def}{=} \begin{cases} deadline(leftwing(h)) \\ \quad \text{, if } leftwing(h) \neq -1 \\ deadline^{L4L}(h) \\ \quad \text{, otherwise} \end{cases} \tag{6.97}$$

A remarkable point to note is that in the case $leftwing(h) = -1$ the lower bound is not necessarily the value one as it would be reasonable to expect, because the demand could be left shifted to period 1 without violating the earliest due date rule. The explanation is that shifting demands for an item to the left should help to build large lots for that particular item. Left shifts can therefore be omitted if there is no other demand for that item in earlier periods, or, which is a bit more restrictive, if there is no demand for which the earliest due date rule must be respected. The lower bound $deadline^{L4L}(h)$ which equals the initial deadline entry of node h prevents a left shift of the demand represented by node h.

What is done for left shift operations can also be done for right shift operations with minor modifications. Due to the earliest due date rule we have a precedence relation between each node $h \in \mathcal{L}$ and every node in the set

$$\mathcal{RW}(h) \stackrel{def}{=} \{k \in \mathcal{L} \mid \begin{array}{l} path(k) = path(h) \wedge \\ deadline^{L4L}(k) > deadline^{L4L}(h)\} \end{array} \tag{6.98}$$

or, which is more efficient, between each node $h \in \mathcal{L}$ and the node $rightwing(h)$ where

$$rightwing(h) \stackrel{def}{=} \begin{cases} \min(\mathcal{RW}(h)) & \text{, if } \mathcal{RW}(h) \neq \emptyset \\ -1 & \text{, otherwise} \end{cases} \tag{6.99}$$

and -1 again signals that there is no additional precedence relation due to the earliest due date rule. Note, for $h \in \mathcal{L}$ the following properties identify the *rightwing* and the *leftwing* function being reversal:

$$h = rightwing(leftwing(h)), \text{ if } leftwing(h) \neq -1 \tag{6.100}$$

$$h = leftwing(rightwing(h)), \text{ if } rightwing(h) \neq -1 \tag{6.101}$$

For $h \in \mathcal{L}$ an upper bound for the deadline entry is

$$deadline^{II}_{UB}(h) \stackrel{def}{=} \begin{cases} deadline(rightwing(h)) \\ \qquad \text{, if } rightwing(h) \neq -1 \\ deadline^{L4L}(h) \\ \qquad \text{, otherwise} \end{cases} . \tag{6.102}$$

As a result we have that shifting demand to the left or to the right by decreasing or increasing the deadline value of a node

$h \in \mathcal{L}$ should be done so that the new entry $deadline(h)$ fulfills

$$\begin{array}{c} \max\{deadline^{I}_{LB}(h), deadline^{II}_{LB}(h)\} \\ \leq \\ deadline(h) \\ \leq \\ \min\{deadline^{I}_{UB}(h), deadline^{II}_{UB}(h)\}. \end{array} \tag{6.103}$$

Note, once the deadline entry of a node h is modified, not only the bounds for the deadline entry of the node h but also the bounds $deadline^{I}_{LB}(succ(h))$, $deadline^{II}_{LB}(rightwing(h))$ (if $rightwing(h) \neq -1$), $deadline^{I}_{UB}(k)$ for $k \in preds(h)$, and $deadline^{II}_{UB}(left\text{–}wing(h))$ (if $leftwing(h) \neq -1$) for deadline entries must be updated before another shift operation can be performed. All other bounds are kept unchanged.

In preliminary tests it turned out that it is best to perform full–size shifts. A left shift of the demand represented by a node h then yields the value

$$deadline(h) = \max\{deadline^{I}_{LB}(h), deadline^{II}_{LB}(h)\}. \tag{6.104}$$

In the case of a right shift, we get

$$deadline(h) = \min\{deadline^{I}_{UB}(h), deadline^{II}_{UB}(h)\}. \tag{6.105}$$

Let us now return to discuss how to select a node to modify its deadline entry. Assume the existence of at least one node $h \in \mathcal{L}$ with $leftwing(h) \neq -1$. Otherwise, demands cannot be shifted and the demand shuffle heuristic repeatedly passes the construction phase using the initial data structure again and again. Hence, consider the set of valid labels

$$\mathcal{VL} \stackrel{def}{=} \{h \in \mathcal{L} \mid leftwing(h) \neq -1\} \tag{6.106}$$

which identifies nodes that are allowed to be shifted. Attached to each node $h \in \mathcal{VL}$ is a priority value π_h which guides the choice of nodes. It is a random process that works as follows. First, choose a label $h^{(0)} \in \mathcal{VL}$ with uniform distribution. Then, iterate the following lines where the index $i = 1, 2, \ldots$ denotes the number of the iteration performed. Choose $h^{(i)} \in \mathcal{VL}$ with uniform distribution. If $\pi_{h^{(i)}} > \pi_{h^{(i-1)}}$ then start a new iteration. Otherwise, terminate the loop with node $h^{(i-1)}$ being the one whose deadline entry should be modified. Note, this process is guaranteed to terminate, because the priority values increase from iteration to iteration. The priority rule that is used in our tests compares the additional holding costs that are incurred when some demand is shifted to the left with the setup costs that are saved when a lot can be built. More formally, we use

$$\pi_h \stackrel{def}{=} \frac{demand(h) h_{item(h)}}{s_{item(h)}} \cdot (deadline(h) - deadline(leftwing(h)) + 1) \tag{6.107}$$

for $h \in \mathcal{VL}$.

Suppose now that we have chosen a node, say $h \in \mathcal{VL}$. The next decision to be made is whether the demand represented by that node should be shifted to the left or to the right. Basically, we do a random choice again where π_{left} and $\pi_{right} = 1 - \pi_{left}$, respectively, are probability values for doing a left or a right shift. Before we give a formal definition of these probabilities, let us motivate some properties. Remember that we perform full-size shift operations, i.e.

$$\Delta t_{left} \stackrel{def}{=} deadline(h) - \max\{deadline^{I}_{LB}(h), deadline^{II}_{LB}(h)\} \tag{6.108}$$

is the number of periods a demand would be shifted to the left, and

$$\Delta t_{right} \stackrel{def}{=} \min\{deadline^{I}_{UB}(h), deadline^{II}_{UB}(h)\} - deadline(h) \tag{6.109}$$

is the number of periods a demand would be shifted to the right. Note, $\Delta t_{left} \geq 0$ and $\Delta t_{right} \geq 0$ remains true. To ease the notation, let $Deadline(h)$ denote the period in which the production that meets the (internal or external) demand represented by node h is actually needed, i.e.

$$Deadline(h) \stackrel{def}{=} \begin{cases} deadline(succ(h)) & , \text{if } succ(h) \neq -1 \\ deadline^{L4L}(h) & , \text{otherwise} \end{cases} \tag{6.110}$$

To start with, assume that $\Delta t_{left} + \Delta t_{right} > 0$ holds. This is the case in which the demand of the selected node h can indeed be shifted to some direction. The conditions

$$\Delta t_{left} = 0 \Rightarrow \pi_{left} = 0 \tag{6.111}$$

and

$$\Delta t_{right} = 0 \Rightarrow \pi_{right} = 0 \tag{6.112}$$

then should hold to enforce that the demand is shifted if it is possible to do so. Another relevant point is that left shift operations cause additional holding costs and thus right shifts should be preferred if Δt_{right} exceeds Δt_{left}. Mathematically, this means to fulfill the condition

$$\Delta t_{right} \geq \Delta t_{left} \Rightarrow \pi_{right} > \pi_{left}. \tag{6.113}$$

Furthermore, a key idea of the shifting of demands is to facilitate the building of lots, hence a large value Δt_{left} or Δt_{right}, respectively, should result in a high probability π_{left} or π_{right}. Eventually, the importance of deciding for a left or a right shift decreases as the difference $Deadline(h) - deadline(h)$ increases. A definition of π_{left} (and π_{right}) that meets all the points mentioned above is

$$\pi_{left} \stackrel{def}{=} \frac{1}{1 + \frac{\sum_{\Delta t=1}^{\Delta t_{right}} \frac{1}{Deadline(h) - deadline(h) - \Delta t + 1}}{\sum_{\Delta t=1}^{\Delta t_{left}} \frac{1}{Deadline(h) - deadline(h) + \Delta t + 1}}}, \tag{6.114}$$

if $\Delta t_{left} > 0$, and $\pi_{left} = 0$, otherwise.

The case $\Delta t_{left} + \Delta t_{right} = 0$ virtually means that the demand of the selected node cannot be shifted to any direction. But we do not give up soon and decide if we would like to shift that demand to the left or to the right by means of a random choice again. Assume that we decided for a left shift. We then perform recursive full–size left shift operations with all preceding nodes in the gozinto–tree starting with the leafs of the tree. Afterwards, evaluating Δt_{left} again may give a positive value. If this is so, we shift the demand to the left and we are done. If Δt_{left} still is zero, we keep the left shifted internal demand at its new position, but do not modify the deadline entry of the original node h. We proceed analogously when a right shift operation should be performed. This time, the successor nodes are recursively shifted to the right starting with the root node of the gozinto–tree containing node h. If we end up with a positive value Δt_{right}, the demand represented by node h is right shifted, too. Otherwise, we keep the changes but do not shift the original demand. What is left, is a definition of the probabilities π_{left} (and π_{right}) with which we decide for a left or a right shift, respectively. In contrast to the definition above, we now use

$$\pi_{left} \stackrel{def}{=} \frac{1}{1 + \frac{(deadline^{L4L}(h) - deadline(h)) demand(h) h_{item(h)}}{s_{item(h)}}}. \qquad (6.115)$$

---

if (at least one feasible solution has been found
during former iterations)
repeat $SHIFTOPS$ times
choose $h \in \mathcal{VL}$.
decide whether h should be shifted to the left or
to the right.
if (h should be shifted to the left)
$leftshift(h)$.
else
$rightshift(h)$.

Table 6.43: The Data Structure Manipulation

This formula is chosen so that we face a fifty–fifty chance for a left shift or a right shift, respectively, if the costs for holding the quantity $demand(h)$ in inventory up to the initial period $deadline^{L4L}(h)$ equal the setup costs for $item(h)$. The greater the difference between the node's current deadline and its initial deadline, the more probable is a right shift of that demand.

In summary, Table 6.43 describes the manipulation of the data structure as it happens after each construction phase. The parameter $SHIFTOPS$ is specified by the user and determines the number of shift operations to be executed. Note, due to the definition of the initial data structure, the very first data structure manipulations actually are left shift operations only. Since shifting demand to the left bears the risk of infeasibility, we only enter the data structure manipulation phase if at least one feasible solution has been found in previous construction phases.

Table 6.44 gives a precise definition of how to shift the demand represented by a node $h \in \mathcal{VL}$ to the left.

$$\Delta t_{left} := deadline(h) - \max\{deadline^{I}_{LB}(h), deadline^{II}_{LB}(h)\}.$$

if $(\Delta t_{left} = 0)$

$\quad treeleftshift(h)$.

else

$\quad$ if $\left(\forall deadline(h) - \Delta t_{left} \leq t < deadline(h) : \sum_{\tau=1}^{t} C_{m_{item(h)}\tau} \geq p_{item(h)} demand(h) + \sum_{\substack{k \in \mathcal{L} \\ deadline(k) \leq t \\ m_{item(k)} = m_{item(h)}}} p_{item(k)} demand(k) \right)$

$\quad\quad deadline(h) := \max\{deadline^{I}_{LB}(h), deadline^{II}_{LB}(h)\}$.

Table 6.44: The Left Shift Procedure $leftshift(h)$

Table 6.45 provides the details of shifting the demand represented by node h and all the internal demand for fulfilling it to the left.

Similarly, Table 6.46 provides the details of how to shift the demand represented by a node $h \in \mathcal{VL}$ to the right.

See Table 6.47 for a specification of shifting the demand represented by node h and all its successors to the right.

6.7.4 Modifications of the Construction Scheme

The construction scheme presented in Section 6.2 needs to be modified in order to generate a production plan which respects the

---

```
if (preds(h) ≠ ∅)
      for k ∈ preds(h)
            treeleftshift(k).
```

$\Delta t_{left} := deadline(h)$
$\quad - \max\{deadline^{I}_{LB}(h), deadline^{II}_{LB}(h)\}.$

if $(\Delta t_{left} \neq 0)$

$\quad$ if ($\forall deadline(h) - \Delta t_{left} \leq t < deadline(h)$:

$$\sum_{\tau=1}^{t} C_{m_{item(h)}\tau} \geq p_{item(h)} demand(h) + \sum_{\substack{k \in \mathcal{L} \\ deadline(k) \leq t \\ m_{item(k)} = m_{item(h)}}} p_{item(k)} demand(k))$$

$\qquad deadline(h) := \max\{deadline^{I}_{LB}(h), deadline^{II}_{LB}(h)\}.$

Table 6.45: The Left Shift Procedure $treeleftshift(h)$

$\Delta t_{right} := \min\{deadline^{I}_{UB}(h), deadline^{II}_{UB}(h)\}$
$\quad - deadline(h).$

if $(\Delta t_{right} = 0)$

$\quad treerightshift(h).$

else

$\quad deadline(h) := \min\{deadline^{I}_{UB}(h), deadline^{II}_{UB}(h)\}.$

Table 6.46: The Right Shift Procedure $rightshift(h)$

upper bounds for the production quantities. We now define a modified version *DS–construct* of the original construction scheme. All auxiliary variables are initialized as before. Additionally, we have upper bounds Q_{jt} for $j = 1, \ldots, J$ and $t = 1, \ldots, T$ where $Q_{j(T+1)} = 0$ for $j = 1, \ldots, J$ is assumed for the sake of notational convenience.

Starting with the program given in Table 6.48 we again begin with fixing the setup states of the machines at the end of period T. This piece of code is essentially the same as the one given in Table 6.1.

```
if (succ(h) ≠ −1)
      treerightshift(succ(h)).
Δt_right := min{deadline^I_UB(h), deadline^II_UB(h)}
            −deadline(h).
if (Δt_right ≠ 0)
      deadline(h) := min{deadline^I_UB(h), deadline^II_UB(h)}.
```

Table 6.47: The Right Shift Procedure *treerightshift*(h)

```
choose j_mT ∈ I_mT.
if (j_mT ≠ 0)
      y_{j_mT T} := 1.
if (m = M)
      DS-construct(T, 0, 1).
else
      DS-construct(T, 1, m + 1).
```

Table 6.48: Evaluating *DS–construct*$(T, 1, \cdot)$

A definition of $\mathcal{I}_{mt}$ and a rule to choose j_{mt} out of it will be given later. After all machines are assigned certain setup states at the end of period t, the procedure given in Table 6.49 takes over.

Note, the production quantity $q_{j_{mt}t}$ is not only constrained by the available capacity and the cumulative demand for item j_{mt} but also by the upper bound $Q_{j_{mt}t}$ which is derived from the data structure Γ^{DS} (see also (6.90)).

Table 6.50 specifies how to evaluate calls of the form *DS–construct*$(t, 1, \cdot)$ where $1 < t < T$. In the case that *DS–construct* $(1, 1, \cdot)$ is called, we execute the code given in Table 6.4.

An important point is the part which tries to increase the capacity utilization in period $t+1$. Table 6.51 provides more details.[14]

[14]For the sake of notational convenience, let us assume that $CD_{0(t+1)} = 0$, $Q_{0(t+1)} = 0$, $q_{0(t+1)} = 0$, and $p_0 = 0$. Testing whether or not $j_{mt} \neq 0$ and $j_{m(t+1)} \neq 0$ to ensure well–defined variable accesses can then be left out.

for $j \in \mathcal{J}_m$

$CD_{jt} := \min\left\{CD_{j(t+1)} + \tilde{d}_{jt}, \max\{0, nr_j - \sum_{\tau=t+1}^{T} q_{j\tau}\}\right\}.$

$Q_{jt} := \min\{\ Q_{j(t+1)} + \sum_{\substack{h\in\mathcal{L}\\ item(h)=j\\ deadline(h)=t}} demand(h),$

$\max\{0, nr_j - \sum_{\tau=t+1}^{T} q_{j\tau}\}\}.$

if $(j_{mt} \neq 0)$

$q_{j_{mt}t} := \min\left\{CD_{j_{mt}t}, Q_{j_{mt}t}, \frac{RC_{mt}}{p_{j_{mt}}}\right\}.$

$Q_{j_{mt}t} := Q_{j_{mt}t} - q_{j_{mt}t}.$

$CD_{j_{mt}t} := CD_{j_{mt}t} - q_{j_{mt}t}.$

$RC_{mt} := RC_{mt} - p_{j_{mt}} q_{j_{mt}t}.$

for $i \in \mathcal{P}_{j_{mt}}$

if $(t - v_i > 0$ and $q_{j_{mt}t} > 0)$

$\tilde{d}_{i(t-v_i)} := \tilde{d}_{i(t-v_i)} + a_{ij_{mt}} q_{j_{mt}t}.$

if $(m = M)$

DS–construct$(t-1, 1, 1)$.

else

DS–construct$(t, 0, m+1)$.

Table 6.49: Evaluating *DS–construct*$(t, 0, \cdot)$ where $1 \leq t \leq T$

---

choose $j_{mt} \in \mathcal{I}_{mt}$.
if $(j_{mt} \neq 0)$
    $y_{j_{mt}t} := 1$.
    if $(j_{mt} \neq j_{m(t+1)})$
        $q_{j_{mt}(t+1)} := \min\left\{CD_{j_{mt}(t+1)}, Q_{j_{mt}(t+1)}, \frac{RC_{m(t+1)}}{p_{j_{mt}}}\right\}$.
        $Q_{j_{mt}(t+1)} := Q_{j_{mt}(t+1)} - q_{j_{mt}(t+1)}$.
        $CD_{j_{mt}(t+1)} := CD_{j_{mt}(t+1)} - q_{j_{mt}(t+1)}$.
        $RC_{m(t+1)} := RC_{m(t+1)} - p_{j_{mt}} q_{j_{mt}(t+1)}$.
        for $i \in \mathcal{P}_{j_{mt}}$
            if $(t + 1 - v_i > 0$ and $q_{j_{mt}(t+1)} > 0)$
                $\tilde{d}_{i(t+1-v_i)} := \tilde{d}_{i(t+1-v_i)} + a_{ij_{mt}} q_{j_{mt}(t+1)}$.
try to increase the capacity utilization in period $t + 1$.
if $(m = M)$
    DS–construct$(t, 0, 1)$.
else
    DS–construct$(t, 1, m + 1)$.

Table 6.50: Evaluating *DS–construct*$(t, 1, \cdot)$ where $1 \leq t < T$

$\hat{t} := t.$

while ($(CD_{j_{mt}(t+1)} > Q_{j_{mt}(t+1)}$ or $CD_{j_{m(t+1)}(t+1)} > Q_{j_{m(t+1)}(t+1)})$
  and $RC_{m(t+1)} > 0$ and $\hat{t} \geq 1)$

  if $(\exists h \in \mathcal{L} : \; (item(h) = j_{mt} \vee item(h) = j_{m(t+1)})$
    $\wedge deadline(h) = \hat{t}$
    $\wedge \min\{ \; deadline^{I}_{UB}(h), \; deadline^{II}_{UB}(h)\} \geq t+1$
    $\wedge CD_{item(h)(t+1)} > Q_{item(h)(t+1)})$

    $deadline(h) := t+1.$

    $Q_{item(h)(t+1)} := \min\{ \; Q_{item(h)(t+1)} + demand(h), \; \max\{0, nr_{item(h)} - \sum_{\tau=t+1}^{T} q_{item(h)\tau}\}\}.$

    $\Delta q_{item(h)(t+1)} := \min\{ \; CD_{item(h)(t+1)}, Q_{item(h)(t+1)}, \frac{RC_{m(t+1)}}{p_{item(h)}}\}.$

    $q_{item(h)(t+1)} := q_{item(h)(t+1)} + \Delta q_{item(h)(t+1)}.$

    $Q_{item(h)(t+1)} := Q_{item(h)(t+1)} - \Delta q_{item(h)(t+1)}.$

    $CD_{item(h)(t+1)} := CD_{item(h)(t+1)} - \Delta q_{item(h)(t+1)}.$

    $RC_{m(t+1)} := RC_{m(t+1)} - p_{item(h)} \Delta q_{item(h)(t+1)}.$

    for $i \in \mathcal{P}_{item(h)}$

      if $(t+1-v_i > 0$ and $\Delta q_{item(h)(t+1)} > 0)$

        $\tilde{d}_{i(t+1-v_i)} := \; \tilde{d}_{i(t+1-v_i)} + a_{i(item(h))} \Delta q_{item(h)(t+1)}.$

  else

    $\hat{t} := \hat{t} - 1.$

Table 6.51: A Procedure to Increase the Capacity Utilization in Period $t+1$

The idea behind this piece of code is the following: Suppose, we have scheduled two items j_{mt} and $j_{m(t+1)}$ on machine m in period $t+1$ where $q_{j_{mt}(t+1)}$ and $q_{j_{m(t+1)}(t+1)}$, respectively, are the corresponding production quantities. It may then happen that the remaining capacity $RC_{m(t+1)}$ of machine m is positive. Furthermore, assume that in period $t+1$ the cumulative demand for item j_{mt}, or $j_{m(t+1)}$, or both is not totally met. Such situations may occur due to left shift operations which have modified the initial data structure. Since the machine is now set up for these items anyhow, there seems to be no reason why we should not use the capacity that is left over in period $t+1$ to fulfill some additional demand. For the reason of consistency, we try to manipulate the data structure in order to shift some demand back to the right. To do so, we scan the data structure for nodes that represent demand for these two items and, if feasible, perform right shifts to make additional production quantities $\Delta q_{j_{mt}(t+1)}$ or $\Delta q_{j_{m(t+1)}(t+1)}$, respectively, available in period $t+1$ until all cumulative demand is fulfilled or until no free capacity remains. As a side effect of increasing the deadline entry of a node, the upper bounds for the deadline entries of some other nodes may change giving the opportunity to set their deadline value to $t+1$, too. Thus, we consider the nodes in a certain order. That is, nodes with a high current *deadline*–value are tested before nodes with a low value are, because shifting the demand that is represented by the latter ones to the right does not have an impact on whether or not demand that is represented by the former ones can be right shifted. The auxiliary variable $\hat{t}$ is used to implement this aspect. However, this defines only a partial order among the nodes. In our implementation, remaining ties are broken by favoring nodes with a high label.

The open question that remains to be discussed is the way an item j_{mt} is chosen out of $\mathcal{I}_{mt}$, and $\mathcal{I}_{mt}$ itself needs to be defined, of course (see Tables 6.48 and 6.50). The set of items machine $m = 1, \ldots, M$ may be set up for at the end of period $t = 1, \ldots, T$ can be defined as the set of items with demand in period t or $t+1$, and with some production being allowed regarding the information in the data structure. More formally,

$$\begin{aligned}\mathcal{I}_{mt} \stackrel{def}{=} \{ & j \in \mathcal{J}_m \mid CD_{j(t+1)} + \tilde{d}_{jt} > 0\} \\ & \cap \{j \in \mathcal{J}_m \mid \; Q_{j(t+1)} > 0 \\ & \qquad \vee \exists h \in \mathcal{L} : (\; item(h) = j \\ & \qquad\qquad \wedge deadline(h) = t)\}.\end{aligned} \tag{6.116}$$

As before, we choose $j_{mt} = 0$, if $\mathcal{I}_{mt} = \emptyset$, and again we do not allow lot splitting in our tests.

For the selection of an item $j_{mt} \in \mathcal{I}_{mt}$ we have several alternatives, some of them are already presented in earlier sections. In the tests presented here, we make a random choice where the probability to choose an item $j \in \mathcal{I}_{mt}$ is defined on the basis of a priority value

$$\pi_{jt} \stackrel{def}{=} h_j Q_{j(t+1)} \tag{6.117}$$

which is an easy–to–compute estimate of the holding costs that are charged if item j is not scheduled in period $t+1$. If

$$\sum_{j \in \mathcal{I}_{mt}} \pi_{jt} = 0$$

a random choice with uniform distribution is done. Otherwise,

$$\frac{\pi_{jt}}{\sum_{i \in \mathcal{I}_{mt}} \pi_{it}}$$

is the probability to choose j out of $\mathcal{I}_{mt}$.

6.7.5 An Example

Consider the example given in Subsection 6.3.5 again. Suppose, the data structure Γ^{DS} is given to be

$$\begin{aligned}\Gamma^{DS} = \{ & (1,7,7,20,0,1,-1, \quad \{(2,6,6,20,1,7,0, \\ & \qquad\qquad\qquad\qquad\qquad \{(4,5,5,20,2,39,1,\emptyset)\}), \\ & \qquad\qquad\qquad\qquad (3,6,6,20,3,8,0, \\ & \qquad\qquad\qquad\qquad\qquad \{(4,5,5,20,4,44,3,\emptyset)\})\}) \\ & (4,8,8,5,5,4,-1,\emptyset),\end{aligned}$$

$$
\begin{aligned}
(1,10,10,20,6,1,-1,\{\ &(2,9,9,20,7,7,6,\\
&\qquad\qquad \{(4,8,8,20,8,39,7,\emptyset)\}),\\
&(3,6,9,20,9,8,6,\\
&\qquad\qquad \{(4,5,8,20,10,44,9,\emptyset)\})\})
\end{aligned}
$$

$$\}$$

which can be illustrated as in Table 6.52 where an entry of the form $L : D^{LB/UB}$ in row j and column t depicts D units of demand for item j that is to be met no later than period t. The node in the data structure that represents this demand has the label L. If the data structure is manipulated, this demand cannot be shifted to a period prior to period LB or after period UB.

Γ^{DS} $t =$	$\ldots 5$	6	7	8	9	10
$j = 1$			$0 : 20^{7/7}$			$6 : 20^{10/10}$
$j = 2$		$1 : 20^{6/6}$			$7 : 20^{9/9}$	
$j = 3$		$3 : 20^{6/6}$				
		$9 : 20^{6/9}$				
$j = 4$	$2 : 20^{5/5}$			$5 : 5^{8/8}$		
	$4 : 20^{5/5}$			$8 : 20^{5/8}$		
	$10 : 20^{5/5}$					

Table 6.52: A Data–Structure for the Demand Shuffle Heuristic

For instance, let us have a look at the node with label 9. It represents a demand of $demand(9) = 20$ units of item $item(9) = 3$ which is to be met no later than period $deadline(9) = 6$. Furthermore, the information in the data structure tells us that this demand is an internal demand caused by producing the demand represented by node $succ(9) = 6$ which is external demand, because $succ(6) = -1$. In turn, fulfilling the demand represented by node 9 causes internal demand that is represented by the nodes in the set $preds(9) = \{10\}$. Hence, the path from the root node of the gozinto–tree to the node 10 contains the items $item(6) = 1$, $item(9) = 3$, and $item(10) = 4$. The position of the node 10 is thus uniquely identified by $path(10) = (\underline{1} \cdot 5 + \underline{3}) \cdot 5 + \underline{4} = 44$ which equals the position of node 4. An additional shift operation may move the demand of node 9 to no period prior to period

$$\begin{aligned}&\max\{deadline^{I}_{LB}(9), deadline^{II}_{LB}(9)\}\\&= \max\{deadline(10) + v_4, deadline(3)\}\\&= 6\end{aligned}$$

and to no period later than period

$$\begin{aligned}&\min\{deadline^{I}_{UB}(9), deadline^{II}_{UB}(9)\}\\&= \min\{deadline(6) - v_1, deadline^{L4L}(9)\}\\&= 9.\end{aligned}$$

Although it appears to be that the demand represented by the node 8 could be left shifted to period 5 next, we would not do so, because the condition (6.91) would no longer be valid for period 5 and machine 1, i.e.

$$\begin{aligned}&\sum_{\tau=1}^{5} C_{1\tau}\\&= 75 \not\geq 80\\&= p_{item(8)}demand(8) + \sum_{k\in\{2,4,10\}} p_{item(k)}demand(k),\end{aligned}$$

which indicates that no feasible solution can be found.

Table 6.53 provides the first few steps of a run of the construction scheme when using the data structure given above.

A possible outcome of the construction phase is shown in Figure 6.6.

Some points of interest are worth to be highlighted:

Step 1: $1 \in \mathcal{I}_{1T}$, because $\tilde{d}_{1T} > 0$, $deadline(6) = 10$, and $item(6) = 1$.

Step 6: Although $\tilde{d}_{39} > 0$ holds, item 3 is not contained in the set $\mathcal{I}_{29}$. This is because $Q_{3T} = 0$ and there is no node h in the data structure which fulfills $deadline(h) = 9$ and $item(h) = 3$. In other words, the demand for item 3 was shifted to the left.

Step 13: We cannot choose any other item than item 4 unless we allow lot splitting which we do not here. Suppose that lot splitting would be allowed, we would have both, item 1 and item 4 in $\mathcal{I}_{17}$.

Step	$(t, \Delta t, m)$	$\vec{\tilde{d}}_{jt}$	$\vec{CD}_{j(t+\Delta t)}$	$\vec{Q}_{j(t+\Delta t)}$	$\mathcal{I}_{mt}$	j_{mt}	$q_{j_{mt}(t+\Delta t)}$
1	(10,1,1)	(20,0,0,0)	(0,0,0,0)	(0,0,0,0)	{1}	1	
2	(10,1,2)	(20,0,0,0)	(0,0,0,0)	(0,0,0,0)	$\emptyset$	0	
3	(10,0,1)	(20,0,0,0)	(20,0,0,0)	(20,0,0,0)			$q_{1T} = 15$
4	(10,0,2)	(20,0,0,0)	(5,0,0,0)	(5,0,0,0)			
5	(9,1,1)	(0,15,15,0)	(5,0,0,0)	(5,0,0,0)	{1}	1	
6	(9,1,2)	(0,15,15,0)	(5,0,0,0)	(5,0,0,0)	{2}	2	$q_{2T} = 0$
7	(9,0,1)	(0,15,15,0)	(5,0,0,0)	(5,0,0,0)			$q_{19} = 5$
8	(9,0,2)	(0,15,15,0)	(0,15,15,0)	(0,20,0,0)			$q_{29} = 15$
9	(8,1,1)	(0,5,5,20)	(0,0,15,0)	(0,5,0,0)	{4}	4	$q_{49} = 0$
10	(8,1,2)	(0,5,5,20)	(0,0,15,0)	(0,5,0,0)	{2}	2	
11	(8,0,1)	(0,5,5,20)	(0,0,15,20)	(0,5,0,25)			$q_{48} = 15$
12	(8,0,2)	(0,5,5,20)	(0,5,20,5)	(0,5,0,10)			$q_{28} = 5$
13	(7,1,1)	(20,0,0,5)	(0,0,20,5)	(0,0,0,10)	{4}	4	
14	(7,1,2)	(20,0,0,5)	(0,0,20,5)	(0,0,0,10)	$\emptyset$	0	
...							

Table 6.53: A Protocol of the Construction Scheme of the Demand Shuffle Procedure

A case in which $Q_{j(t+1)} > 0$ and $CD_{j(t+1)} = 0$ holds for an item j and a period t does not occur in this example. It is, however, remarkable to note that we may face such situations. Assume for instance that there would not be any external demand for item 4 and that all (internal) demand is shifted leftmost (let us suppose this could be done without violating the capacity constraints (6.91)). Then, in Step 9, the set $\mathcal{I}_{18}$ would be empty. That is, the presented procedure would handle such cases correctly.

6.7.6 Experimental Evaluation

The demand shuffle procedure is tested on the basis of the small PLSP-MM-instances given in Section 4.2. We perform 1,000 iterations and choose the method parameter $SHIFTOPS = 10$. Tables B.13 to B.15 in the Appendix B present the details of the computational study. Again, the data given there contain for each

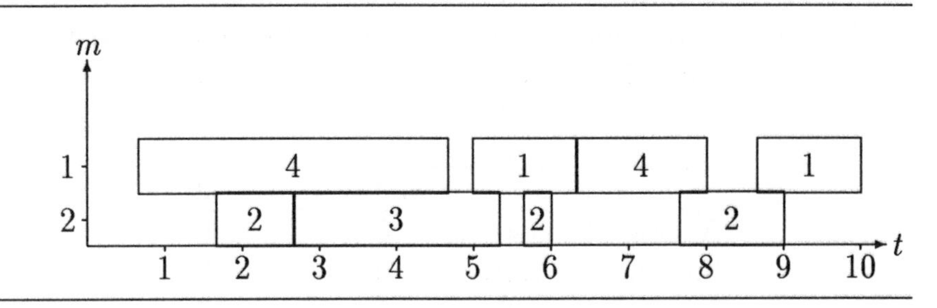

Figure 6.6: A Possible Outcome of the Run in the Protocol

parameter level combination the average deviation from the optimum objective function value, the worst case deviation, and the number of instances for which a feasible solution is found. See (6.25) for a definition of the deviation.

The effect of changing certain parameter levels is once more analyzed on the basis of aggregated data. Table 6.54 shows what happens if the number of machines is varied. Only slight differences can be measured for the average deviation from the optimum. Remarkable to note is the infeasibility ratio. For all instances with $M = 1$ a feasible solution is found and only 1.16% of the instances with $M = 2$ remain unresolved. The run–time increases if the number of machines is increased.

	$M = 1$	$M = 2$
Average Deviation	6.28	6.92
Infeasibility Ratio	0.00	1.16
Average Run–Time	0.38	0.50

Table 6.54: The Impact of the Number of Machines on the Performance

The complexity of the gozinto–structure and its impact on the performance is studied in Table 6.55. It can be seen that a high complexity gives significantly larger deviations from the optimum than a low complexity. The infeasibility ratio, however, is almost unaffected. Furthermore, the more complex the gozinto–structures are, the more run–time is needed to solve the instances.

	$\mathcal{C} = 0.2$	$\mathcal{C} = 0.8$
Average Deviation	5.43	7.80
Infeasibility Ratio	0.57	0.59
Average Run–Time	0.41	0.47

Table 6.55: The Impact of the Gozinto–Structure Complexity on the Performance

Table 6.56 presents different results for different demand patterns. Though the average deviation from the optimum varies among the parameter levels, there is no clear tendency which could prove that sparsely–filled demand matrices give lower or higher average deviations than matrices with many non–zeroes. With respect to the run–time, sparsely–filled demand matrices give best results. The infeasibility ratio shows that only for $(T_{macro}, T_{micro}, T_{idle}) = (5, 2, 2)$ some instances remain unresolved.

	$(T_{macro}, T_{micro}, T_{idle}) =$		
	$(10, 1, 5)$	$(5, 2, 2)$	$(1, 10, 0)$
Average Deviation	7.07	5.80	6.83
Infeasibility Ratio	0.00	1.85	0.00
Average Run–Time	0.53	0.50	0.30

Table 6.56: The Impact of the Demand Pattern on the Performance

The effect of varying the cost structure is analyzed in Table 6.57. If the ratio of setup and holding costs is high, the demand shuffle method performs best in terms of the average deviation from the optimum. The infeasibility ratio is unaffected by changes in the cost structure. Surprisingly, the average run–time declines if setup costs grow.

	$COSTRATIO =$		
	5	150	900
Average Deviation	9.68	5.25	4.85
Infeasibility Ratio	0.58	0.58	0.58
Average Run–Time	0.46	0.45	0.40

Table 6.57: The Impact of the Cost Structure on the Performance

To see if the capacity utilization has an impact on the performance, let us have a look at Table 6.58. We see that if the capacity utilization inclines, the average deviation from the optimum does so, too. Those few instances for which no feasible solution is detected, are instances with a high capacity utilization. The run–time is more or less unaffected.

	$U = 30$	$U = 50$	$U = 70$
Average Deviation	5.15	6.76	8.06
Infeasibility Ratio	0.00	0.00	1.88
Average Run–Time	0.45	0.45	0.42

Table 6.58: The Impact of the Capacity Utilization on the Performance

In summary, only six out of the 1,033 instances in the test–bed remain unresolved which is an amazing result. The overall infeasibility ratio is 0.58%. The average run–time is 0.44 CPU–seconds. The overall average deviation from the optimum objective function value is an exciting 6.60%.

Most of the intelligence of the demand shuffle heuristic lies in the way a node is selected for a shift operation, in the way to decide whether a left or a right shift should be performed, and in the way we select a setup state for a machine. To show that the presented rules do indeed make a contribution, Table 6.59 shows what happens if we select a node with uniform distribution and try to shift it then, assume a fifty–fifty chance for a left or a right shift, and select the setup state of a machine with uniform distribution among the valid candidates.

	Average Deviation	Infeasibility Ratio	Average Run–Time
Intelligent Decision Rules	6.60	0.58	0.44
Simple Decision Rules	7.71	0.87	0.40

Table 6.59: The Impact of the Decision Rules on the Performance

In terms of the run–time performance it turns out that we have to pay a cheap price only for using intelligent decision rules. The advantage of these rules is that on the basis of both, the average deviation from the optimum and the infeasibility ratio, a better performance is gained. However, using no intelligence in the demand shuffle method still gives very good results when compared with other heuristics.

6.8 Summary of Evaluation

We have presented five heuristics for the PLSP–MM and performed an in–depth computational study for each of them by trying to solve 1,033 test instances which were systematically generated as described in Section 4.2. Table 6.60 reviews the overall average results for these procedures.

It turns out that solving PLSP–MM–instances is quite a hard task. The demand shuffle procedure decidedly outperforms all other heuristics in terms of the average deviation from the optimum objective function value and in terms of the infeasibility ratio. Second best is the randomized regret based sampling method closely followed by the cellular automaton. The disjunctive arc based tabu search is disappointing. So is the genetic algorithm if only small population sizes are considered. The genetic

	Average Deviation	Infeasibility Ratio	Average Run–Time
Randomized Regret Based Sampling	10.33	9.68	0.45
Cellular Automaton	10.94	12.97	0.54
Genetic Algorithm	19.89	17.52	0.10
Disjunctive Arc Based Tabu Search	17.59	35.62	0.50
Demand Shuffle	6.60	0.58	0.44

Table 6.60: Summary of Heuristic Procedures

algorithm, however, is the fastest available code and its results are improved if large populations are evaluated.

6.9 Some Tests with Large Instances

Up to here, we have tested the heuristics for small instances only. It is, however, of extreme importance if these heuristics can be applied to large instances as well. And if they are, we like to know if the ranking of the heuristics is the same as for the small instances. Especially, the demand shuffle heuristic must proof its dominance for large instances. That is what we like to find out in this section with a preliminary computational study.

The test–bed that we use consists of several parameter level combinations of J and T which are assumed to be an adequate measure for the size of an instance. For each parameter level combination we use the instance generator APCIG described in Chapter 4 to create five instances each. The parameters for the instances generator or chosen as follows:

Gozinto–structures:
$ARC_{min} = 1.$
$ARC_{max} = 1.$
$COMPLEXITY = 0.2.$

Holding and setup costs:
$HCOST_{min} = 1$.
$HCOST_{max} = J$.
$COSTRATIO = 900$.

Number of machines:
$M = \frac{2}{5}J$.

Demand pattern:
$(T_{macro}, T_{micro}, T_{idle}) = (\frac{T}{5}, 5, 2)$.

Capacity utilization:
$U = 50$.

All other parameters are chosen as in Section 4.2.

The values of the method parameters of the five heuristics are exactly the same as for the small PLSP–MM–instances for which the results are reported in Section 6.8.

As a performance measure we consider the average deviation from the demand shuffle result where the deviation is computed using

$$deviation \stackrel{def}{=} 100\frac{UB_H - UB_{DS}}{UB_{DS}} \tag{6.118}$$

with UB_H being the upper bound of an instance that is computed by means of the heuristic $H \in \{DS, RR, CA, GA, TS\}$. The short–hand notation DS stands for demand shuffle, RR for randomized regret based sampling, CA for cellular automaton, GA for genetic algorithm, and TS for disjunctive arc based tabu search. In this definition we use the result of the demand shuffle procedure as a point of reference, because (good) lower bounds are not available for large instances.

Table 6.61 summarizes the results of different heuristics when applied to instances of different size. For each combination of heuristic and parameter levels for J and T this table provides the average deviation from the demand shuffle result. To relate the results for the large instances to those for the small instances, we also give the deviation from the small instance demand shuffle results. The number of instances for which a feasible solution is found is given, too. The demand shuffle method is able to find a feasible solution for every large instance in the test-bed.

J	T	RR	CA	GA	TS
5	10	4.04	4.52	13.14	10.19
5	50	-4.03 [5]	22.52 [5]	24.38 [5]	14.39 [5]
10	50	3.47 [5]	42.37 [5]	31.35 [5]	42.57 [5]
20	50	17.47 [5]	38.92 [1]	29.24 [5]	— [0]
5	100	-13.09 [5]	58.07 [5]	27.68 [5]	18.03 [5]
10	100	-1.38 [5]	38.56 [3]	30.80 [5]	— [0]
20	100	11.22 [5]	34.74 [1]	30.81 [5]	— [0]
5	500	-16.02 [5]	59.86 [5]	25.72 [5]	— [0]
10	500	-4.40 [5]	31.47 [4]	28.72 [5]	— [0]
20	500	13.81 [5]	27.64 [2]	24.26 [5]	— [0]

Table 6.61: Performance for Large Instances

It turns out that only the randomized regret based sampling method and the genetic algorithm also find feasible solutions for all instances in the test-bed. The capability of the disjunctive arc based tabu search method to find feasible solutions for large instances is disastrous. In terms of average deviation from the demand shuffle result, both, the cellular automaton and the genetic algorithm give disappointing results, though the genetic algorithm shows better performance in almost all cases. Remarkable to note, the randomized regret based method reveals itself as a competitive candidate for large instances. Especially for instances with a small number of items, i.e. $J \leq 10$, this method convinces. However, if the number of items is large, i.e. $J > 10$, the results of the demand shuffle procedure cannot be reached.

Chapter 7

Parallel Machines

While we have spent much effort to discuss the PLSP–MM on isolation, we will now outline solution procedures for extensions of the lot sizing and scheduling problem. To begin with, the PLSP–PM is considered in this chapter. The PLSP–MR and the PLSP–PRR are the subjects of interest in the two chapters that will follow.

Fortunately, all methods described thus far have a common basis, i.e. the construction scheme given in Section 6.2. For presenting the fundamental ideas it should therefore be sufficient to confine ourself to the adaption of one method to these extensions instead of revealing all details for all methods. We decide to work with the demand shuffle heuristic, not only because of its computational performance but also because of being the most complex one. Finding modifications of the regret based procedure, the cellular automaton, the genetic algorithm, or the disjunctive arc based tabu search is supposed to be a more or less easy task to do.

Lot sizing and scheduling with parallel machines is almost untreated in the literature. In [Pop93], the single–level DLSP is considered. So it is in [Cam92a, Sal91, SaKrKuWa91]. The ELSP is treated in [Car90]. For lot sizing without scheduling some authors assume parallel machines which, for instance, represent regular time and overtime [Der95, DiBaKaZi92a, Hin95]. Section 7.1 gives an introduction into a closely related field, i.e. job scheduling with parallel machines. Afterwards, in Section 7.2, we turn back to the

PLSP–PM and describe some modifications of the demand shuffle heuristic to handle that case. Some preliminary computational results are reported in Section 7.3.

7.1 Related Topics: Job Scheduling

Job scheduling closely relates to lot sizing and scheduling. An enormous number of job scheduling literature is available ([Bru95, Pin95] are recent textbooks that give an overview of the state–of–the–art). To fulfill a certain job, several operations are to be performed. Roughly speaking, the problem is to find a schedule which defines the sequence and the start times of operations that are to be processed under some side constraints. Some examples for restrictions that may be found in job scheduling applications are precedence relations among the operations, release and/or due dates for jobs, scarce capacities, non–preemptive operations, and not to execute operations belonging to the same job concurrently. Beside the terminology (e.g. external demands are called jobs now), the most important difference between lot sizing and scheduling and job scheduling is the objective function that is considered. Instead of minimizing some cost oriented function, time and capacity oriented objectives such as minimizing the makespan, minimizing the lateness, or minimizing the tardiness are of interest.

In the case of parallel machines, the operations are not assigned to a unique machine in advance. Instead, there is a choice on what machine an operation may be performed. Depending on whether or not the choice of a machine affects the processing time of the operation under concern, three subcases of the parallel machine case are discriminated in the literature. If the processing time of the operations stays the same no matter on what machine the operations are performed, the machines are called identical machines. Two machines, say m_a and m_b, are termed to be uniform, if every operation that can be processed on machine m_a can also be processed on machine m_b, and if the ratio of the processing time on machine m_a and the processing time on machine m_b

turns out to be the same for every operation. It should be clear that identical machines are uniform, too. In the most general case, two non–uniform machines are said to be heterogeneous. Note, the PLSP–PM–model covers all three cases and so will the heuristic that is presented in the next section.

There is a large number of publications that deal with the job scheduling problem with parallel machines. Most of it is confined to the identical machine case. A comprehensive review of the job scheduling literature is out of the scope of this work. We refer to [Bak74, Bla87, BlCeSlWe86, BlDrWe91, Bru95, ChSi90, CoMaMi67, GrLaLeRKa79, HoSa76, Pin95] for more details.

7.2 A Demand Shuffle Adaption

The demand shuffle method described in Section 6.7 cannot be applied to the PLSP–PM without some modifications. While the data structure can be kept as it is, the specification of the data structure manipulation as well as the construction scheme must undergo changes.

Let us start with the data structure manipulation and consider the condition (6.91) which can now be stated as

$$\sum_{\tau=1}^{t} C_{m\tau} \geq \sum_{\substack{h \in \mathcal{L} \\ item(h) \in \mathcal{J}_m \\ deadline(h) \leq t}} p_{item(h)m} demand(h) \tag{7.1}$$

for $m = 1, \ldots, M$ and $t = 1, \ldots, T$. The code given in the Tables 6.44 and 6.45 must be adapted likewise. Note, in contrast to the PLSP–MM a violation of this condition does not necessarily mean that no feasible solution can be found, because the right hand sides of the inequalities are upper bounds rather than exact values of the capacity needs in a feasible plan. Hence, it might happen that even the initial data structure does not fulfill this condition. However, we do not allow left shift operations that would violate these inequalities which is a conservative strategy.

For selecting a node for a shift operation, we now use the priority value

$$\pi_h \stackrel{def}{=} \frac{demand(h) h_{item(h)}}{\max\{s_{item(h)m} \mid m \in \mathcal{M}_{item(h)}\}} \cdot (deadline(h) - deadline(leftwing(h)) + 1) \quad (7.2)$$

for $h \in \mathcal{VL}$ instead of (6.107).

This is all that must be changed in the way the data structure is manipulated. So, let us now discuss the changes to the construction scheme. While all variables are defined as before, *DS–PM–construct* denotes the new procedure. Evaluating *DS–PM–construct*$(T, 1, \cdot)$ equals what is given in Table 6.48 making recursive calls to *DS–PM–construct* instead of *DS–construct*, of course. The modification of the piece of code provided in Table 6.49 is given in Table C.1 in the Appendix C. The idea behind both codes is the same, but we must take into account that $\mid \mathcal{M}_j \mid > 1$ for some (maybe all) items $j = 1, \ldots, J$. And thus, using a loop enumerating $j \in \mathcal{J}_m$ at the beginning of the code would increase some values CD_{jt} and Q_{jt} more than once which would be incorrect.

The changes of what is provided in Table 6.50 are shown in Table C.2 in the Appendix C. Likewise, the piece of code given in Table 6.4 must be modified to get the code that describes how to evaluate a call of the form *DS–PM–construct*$(0, 1, \cdot)$.

The procedure to increase the capacity utilization as given in Table 6.51 needs minor modifications only and is therefore not given again. We simply must use q_{jmt} instead of q_{jt} and p_{jm} instead of p_j. Furthermore, when subtracting production quantities from the net requirement we must not forget that the same item may be scheduled on several machines which is similar to the lines in Table C.1.

The selection of a setup state j_{mt} out of $\mathcal{I}_{mt}$ as to be done in Tables 6.48 and C.2 bases on the definition (6.116) for $\mathcal{I}_{mt}$ in our implementation. The priority value for an item, however, differs. This time, we use

$$\pi_{jt} \stackrel{def}{=} h_j Q_{j(t+1)} \frac{\max\{p_{j\mu} \mid \mu \in \mathcal{M}_j\}}{p_{jm}} \frac{\max\{s_{j\mu} \mid \mu \in \mathcal{M}_j\}}{s_{jm}} \quad (7.3)$$

instead of (6.117) to express that items should have a high preference to be scheduled on machines where capacity consumption and setup costs are low.

7.3 Preliminary Experimental Evaluation

To gain some first insight into the performance of the PLSP–PM demand shuffle method, we generate a very small test–bed using the instance generator APCIG with different parameter level combinations of $COSTRATIO \in \{5, 900\}$ and $U \in \{30, 70\}$. For each parameter level combination five instances are created using $J = 5$, $M = 2$, $COMPLEXITY = 0.2$, $(T_{macro}, T_{micro}, T_{idle}) = (5, 2, 2)$, and $MACHPERITEM = 2$. Similar to what is done in Section 4.2 we choose all the other parameter values.

We let the demand shuffle procedure iterate 1,000 times where $SHIFTOPS = 10$ is used. Since we have not developed any special lower bounding procedure for the PLSP–PM, we use the LP–relaxation of the PLSP–PM–model formulation given in Section 3.4 as a point of reference. This is done, because the instances cannot be solved optimally within reasonable time. Table 7.1 shows the average deviation for each parameter level combination where

$$deviation \stackrel{def}{=} 100 \frac{UB - LB}{LB} \tag{7.4}$$

defines the deviation of the upper bound UB from the lower bound LB.

	$U = 30$	$U = 70$
$COSTRATIO = 5$	40.99	30.52
$COSTRATIO = 900$	319.16	526.00

Table 7.1: Performance of the PLSP–PM–Method

The need for good lower bounds is obvious. While the results for $COSTRATIO = 5$ are within reasonable bounds, the average deviation of the upper bound from the lower bound is beyond promising limits. We cannot say which portion of these large deviations stems from poor lower bounds and which stems from the heuristic. However, the demand shuffle method finds feasible solutions.

Chapter 8

Multiple Resources

In this chapter we deal with the PLSP–MR. Although some lot sizing models are with multiple resources (e.g. [Der95, Hel94, Sta88, Sta94]), for solution procedures other than LP–based techniques the single– or multi–machine case is assumed. In Section 8.1 we give an introduction into the resource constrained project scheduling problem (RCPSP) where multiple resources are standard. Then, in Section 8.2, we show how the demand shuffle heuristic can be modified in order to be able to solve the PLSP–MR. Preliminary computational tests are provided in Section 8.3.

8.1 Related Topics: Project Scheduling

Project scheduling is a generalization of job scheduling and thus links to lot sizing and scheduling as well. A project consists of one or more operations to be performed and among which precedence relations are to be respected. In contrast to job scheduling, each operation may now require multiple scarce resources. The objectives that are considered in the literature are often time oriented again, e.g. minimizing a project's makespan. But cost oriented approaches such as maximizing a project's net present value can also be found. For more details we refer to [BeDoPa81, Dav66, Dav73, Dem92, DoPa77, DoDr91, Her72, IcErZa93, Kol95, Pat84, Spr94].

8.2 A Demand Shuffle Adaption

For attacking the PLSP–MR we use the original data structure and manipulate the data structure as described in Section 7.2. Note, this time a violation of the condition (7.1) signals without any doubt that no feasible solution can be found.

The construction scheme *DS–MR–construct* to be applied, is given in Table D.1 in the Appendix D. Note, *DS–MR–construct* has two parameters only. These are t and Δt which have the same meaning as before. The parameter m that is passed to the *DS–construct*–procedure is not needed here. The auxiliary variables introduced in Section 6.2 and Subsection 6.7.4 are used again and have the same initial values. The decision variables q^B_{jt} and q^E_{jt} are initialized with zero.

When developing methods for the PLSP–MR, we must take into account that setting a resource up for an item j at the end of period t comes along with setting all resources $\mathcal{M}_j$ up for j at the end of period t. This has two consequences. First, scheduling an item j blocks any other item i when $\mathcal{M}_i \cap \mathcal{M}_j \neq \emptyset$, i.e. the items share a common resource. And second, if we would consider the resources in a certain order for fixing their setup states, say resource m_a is always set up before resource m_b is, we would systematically give those items which require m_a and m_b a better chance to be selected than those which require m_b but not m_a, because the latter ones are blocked. Hence, we give up the enumeration principle and do not strictly move from resource 1 to resource M any longer. Instead we use $\mathcal{I}_t$ for $t = 1, \ldots, T$ which replaces $\mathcal{I}_{mt}$ and which is the set of items j_t among which we may choose to set some resources up for. The definition of $\mathcal{I}_t$ follows the idea of (6.116), but also takes blocking into account. In our implementation items are chosen out of $\mathcal{I}_t$ at random where the priority values (6.117) are used to define a probability distribution. Moreover, lots are not split and thus some items are chosen with certainty.

Calls of the form *DS–MR–construct*$(t, 0)$ are evaluated using the code in Table D.2 in the Appendix D.

The most important aspect is that the bottleneck resource determines the production quantity that can be scheduled per period. Table D.3 in the Appendix D describes how to proceed.

The way the setup states are chosen in Table D.3 equals the procedure in Table D.1. For determining the production quantities q_{jt}^{B}, we have to take into account the remaining capacity of several resources. The procedure to increase the capacity utilization in period $t+1$ is adapted from Table 6.51 as shown in Table D.4 in the Appendix D. Two sets of items $\mathcal{SI}_{(t+1)}^{B}$ and $\mathcal{SI}_{(t+1)}^{E}$, respectively, contain those items for which a proper setup state is defined at the beginning and at the end of period $t+1$.

Table D.5 in the Appendix D shows what happens when we reach period 0.

After this procedure terminates, a feasibility check can be done as usual. Remember, the postprocessor must not be executed (see Section 3.5).

8.3 Preliminary Experimental Evaluation

To do some first tests with the PLSP–MR demand shuffle method, we generate a small test–bed using the instance generator APCIG with different parameter level combinations of $COSTRATIO \in \{5, 900\}$ and $U \in \{30, 70\}$. For each parameter level combination five instances are created using $J = 5, M = 2, COMPLEXITY = 0.2$, $(T_{macro}, T_{micro}, T_{idle}) = (5, 2, 2)$, and $MACHPERITEM = 2$. All other parameters are set like it is done in Section 4.2.

While running the demand shuffle heuristic, 1,000 iterations are performed where $SHIFTOPS = 10$. Since there is no special lower bounding procedure for the PLSP–MR available, we use the LP–relaxation of the PLSP–MR–model formulation given in Section 3.5 as a point of reference. Similar to the small PLSP–PM–instances, the instances cannot be solved optimally within reasonable time. Table 8.1 shows the average deviation for each parameter level combination where (7.4) defines the deviation.

As for the PLSP–PM, it becomes clear that we will need good lower bounds before any meaningful statement about the perfor-

	$U = 30$	$U = 70$
$COSTRATIO = 5$	123.25	97.92
$COSTRATIO = 900$	577.01	680.21

Table 8.1: Performance of the PLSP–MR–Method

mance of PLSP–MR–methods can be made. It is promising to note, however, that the demand shuffle heuristic finds feasible solutions for these instances.

Chapter 9

Partially Renewable Resources

In this chapter the PLSP–PRR is tackled. The notion of partially renewable resources is new and thus there exists no literature for lot sizing and scheduling with such resources. Section 9.1 shortly draws the attention to course scheduling which is the application field for which this kind of resource was invented. Then, in Section 9.2 we present the modifications of the demand shuffle heuristic that allow us to solve the PLSP–PRR. Finally, in Section 9.3 some preliminary computational tests are done.

9.1 Related Topics: Course Scheduling

Course scheduling comprises the assignment of lectures to professors and the scheduling of lectures. Since schedules are cyclic it usually is sufficient to find a feasible schedule for one week. A week is naturally subdivided into days and each day is in turn split into small time slots, say half–hours. For each classroom and each time slot we then must find a unique assignment to lectures such that several constraints are fulfilled. Such constraints are for instance the completion constraints for each lecture, the fact that no professor can give two lectures simultaneously, and many more. In [DrJuSa93] a MIP–model–formulation is given where the lecturing time a professor spends per day is a partially renewable resource.

9.2 A Demand Shuffle Adaption

Again, the data structure can be used as it is and the data structure manipulation is almost the same as in Section 6.7. All that changes is that in addition to the capacity check (6.91) the condition

$$\sum_{\tau=1}^{\lceil \frac{t}{L} \rceil} \tilde{C}_{m\tau} \geq \sum_{\substack{h \in \mathcal{L} \\ item(h) \in \tilde{\mathcal{J}}_m \\ \lceil \frac{deadline(h)}{L} \rceil \leq \lceil \frac{t}{L} \rceil}} \tilde{p}_{item(h)m} demand(h) \tag{9.1}$$

must hold for $m = 1, \ldots, \tilde{M}$ and $t = 1, \ldots, T$ which restricts left shift operations.

The construction scheme *DS–PRR–construct* is very much alike *DS–construct* which is defined in Section 6.7. Let $\tilde{RC}_{\mu\tau}$ denote the remaining capacity of the partially renewable resource μ in the time interval τ. Initially, $\tilde{RC}_{\mu\tau} = \tilde{C}_{\mu\tau}$ for $\mu = 1, \ldots, \tilde{M}$ and $\tau = 1, \ldots, \tilde{T}$. The construction phase starts with the code given in Table 6.48 where the only difference is that now recursive calls are made to *DS–PRR–construct* instead of *DS–construct*. Both, the definition of $\mathcal{I}_{mt}$ and the way an item is chosen out of this set is exactly the same as for the multi–machine case. Table E.1 in the Appendix E shows how to evaluate calls of the form *DS–PRR–construct*$(\cdot, 0, \cdot)$ where the remaining capacity of the partially renewable resources affects the computation of production quantities.

The code in Table E.2 in the Appendix E takes over afterwards which is derived from what can be seen in Table 6.50 by modifying the computation of production quantities and by updating the remaining capacity of the partially renewable resources.

The procedure to increase the capacity utilization in period $t + 1$ must be modified, too. Table E.3 in the Appendix E tells how.

If period 0 is reached we must follow the lines in Table 6.4 where recursive calls are made to *DS–PRR–construct*, of course, and production quantities are computed as described in Table E.2.

9.3 Preliminary Experimental Evaluation

To get some preliminary results for the PLSP–PRR demand shuffle method, small test instances are generated using the instance generator APCIG with different parameter level combinations of $U \in \{30, 70\}$ and $\tilde{U} \in \{30, 70\}$. For each parameter level combination five instances are created using $J = 5$, $M = 2$, $COMPLEXITY = 0.2$, $COSTRATIO = 900$, $(T_{macro}, T_{micro}, T_{idle}) = (5, 2, 2)$, $\tilde{M} = 1$, $\tilde{L} = 2$, and $\tilde{T} = 5$. All other parameters are chosen in analogy of Section 4.2.

A total of 1,000 iterations of the demand shuffle procedure is performed with $SHIFTOPS = 10$. No special lower bounding procedure for the PLSP–PRR is available. Hence, we use the LP–relaxation of the PLSP–PRR–model formulation given in Section 3.6 as a point of reference. The average deviation for each parameter level combination is provided in Table 9.1 where (7.4) defines the deviation.

	$U = 30$	$U = 70$
$\tilde{U} = 30$	169.05	328.57
$\tilde{U} = 70$	193.23	309.83

Table 9.1: Performance of the PLSP–PRR–Method

Again, we cannot make any statement as long as we have not established any lower bounds for which we know how good or bad they are.

Chapter 10

Rolling Planning Horizon

All methods described thus far are tailored to construct a production plan for T periods. Since the lifetime of a firm is supposed to last beyond the planning horizon, lot sizing and scheduling is not a single event. A quick and dirty approach to meet that situation would be to plan for the T periods $1, \ldots, T$, to implement that plan, to plan for the next T periods $T+1, \ldots, 2T$, afterwards, and so on. This would make short–term production planning being a process running a lot sizing and scheduling method every T periods. Beside the fact that the final setup state of one production plan defines the initial setup state for the next, these runs would be independent.

In a real–world situation, however, this working principle would not be appropriate for several reasons. The demand for instance appears to be non–deterministic, because some customers may withdraw their orders while others put some additional items on order. Furthermore, a more accurate estimate for some external demand refines early forecast as time goes by, and unexpected events such as machine breakdowns may interrupt and delay production.

So, what usually happens is that planning overlaps. This is to say that starting with a plan for the periods $1, \ldots, T$ the plan for the first, say $\Delta T \geq 1$, periods is implemented and a new plan is then generated for the periods $\Delta T+1, \ldots, \Delta T+T$ which coins the name rolling horizon. In other words, the production in the periods $\Delta T + 1, \ldots, T$ is rescheduled. Note, if $\Delta T < \frac{T}{2}$ some periods are revised more than once.

This point of view reveals the lot sizing and scheduling problem with T periods being a subproblem in a rolling horizon implementation. While the first ΔT periods of the current production plan are implemented, new lot sizes and a new schedule may differ markedly from a former production plan in later periods due to rescheduling. This phenomenon is known as nervousness [Bak81, BaPe79, BlKrMi87, Cam92b, CaJuKr79, HaCl85, KrCaJu83, Mat77, MiDa86, SrBeUd87, Ste75, TeDe95]. Since many proceedings on a shop floor such as cutting, packing, and material handling do heavily interact with the production process and the supply chain management is also affected by production, nervous production plans cause high transaction costs. It is unlikely to find methods which take all relevant aspects into account. Hence, the performance of lot sizing and scheduling methods should not only be evaluated by run–time and objective function values for a fixed horizon, but by (cost and (in–)stability) measures for the performance on a rolling horizon basis, too.

In Section 10.1 we review some more literature for planning in rolling horizon implementations. Stability measures are suggested in Section 10.2. Section 10.3 gives another emphasis on why it is important to take initial inventory into account and discusses a modification of the demand shuffle method for the PLSP–MM with initial inventory.

10.1 Literature Review

The question of how to measure the performance of a lot sizing method when applied on a rolling horizon basis is discussed and studied by several authors. There are two main streams. Some authors consider cost oriented measures while others suggest stability oriented performance measures. If computational studies are done, a plan is generated for the periods $1, \ldots, T, \ldots, \hat{T}$ where $\hat{T}$ is a parameter of the test–bed. Note, this is an approximation, because the result for $\hat{T} \to \infty$ would be of interest.

In [Bak77] the ratio of the objective function value of the implemented plan to the optimum objective function value of the overall problem for the periods $1, \ldots, T, \ldots, \hat{T}$ is considered. This value is greater than or equal to one where a value close to one is desired. Comparing several methods when applied on a rolling horizon basis can be done as in [DBoWaGe82] where different methods are applied to the same instances and the objective function values of the implemented plans are compared.

A measure for instability is presented in [BlKrMi86] where $\Delta T = 1$ is assumed. Instability is expressed as the number of lots in the first period which are to be produced after a rescheduling operation but which were not scheduled before. Changes in the size of the lots are not considered. This measure is extended in [Jen93] where those lots are counted as well which are not scheduled any more but which were before. Also, a measure which computes the changes of lot sizes in the first period is introduced. In both cases, a value close to zero indicates little changes. In [SrBeUd88] it is argued that focusing on the first period only and disregarding changes in the size of the lots is inappropriate. Thus, a measure is proposed which takes lot size changes in all periods into account.

Some other work is devoted to find out in what situations an extension of the planning horizon does not affect the plan in early periods [SrBeUd88]. In some cases, a so–called decision horizon $\check{T} < T$ can be determined. For example, see [BeSmYa87, BeCrPr83, BlKu74, ChMo86, ChMoPr90, EpGoPa69, FeTz94a, FeTz95, KlKu78, KuMo73, LuMo75, Mat91, Zab64]. This means that an optimum solution for every instance with more than $\check{T}$ periods can be found such that the optimum solution for the subproblem consisting of periods $1, \ldots, \check{T}$ is a part of the overall optimum solution. In such a case, $\Delta T = \check{T}$ should be chosen.

The impact of the choice of T on the performance of methods for the Wagner–Whitin problem in a rolling horizon implementation is studied in [Bak77, BlMi80, CaBeKr82, Cha82, MaGu95, Ris90]. Surprisingly, choosing large values for T is not the best choice as one could have expected. A significant effect of forecast errors for demand on the performance is proven in [DBoWaGe82, DBoWa83, WeWh84]. Error bounds for implementing the plan in the first ΔT periods without having any information beyond period T are discussed in [ChHeLe95, DeLe91, FeTz94a, LeDe86].

Noteworthy to mention that the stability problem caused by rescheduling does not occur for lot sizing only, but for scheduling as well (e.g. see [DaKo95, WuStCh93]).

10.2 Stability Measures

To measure the performance of a lot sizing method when used with a rolling planning horizon, assume that a T–period subproblem is solved $n > 0$ times. As a result we get a production plan for the periods $1, \ldots, T, \ldots, \hat{T} = (n-1)\Delta T + T$. Following the lines above, in each run $i = 1, \ldots, n-1$ the plan for the periods $(i-1)\Delta T + 1, \ldots, i\Delta T$ is implemented while the plan for the periods $i\Delta T + 1, \ldots, (i-1)\Delta T + T$ is of a preliminary nature. Let

$$\dot{q}_j^{(i)} \stackrel{def}{=} \sum_{t=1}^{T-\Delta T} \zeta_{jt} q_{j(t+i\Delta T)} \tag{10.1}$$

for $i = 1, \ldots, n-1$ denote the weighted production quantities for item $j = 1, \ldots, J$ that are temporarily scheduled and let

$$\ddot{q}_j^{(i)} \stackrel{def}{=} \sum_{t=1}^{T-\Delta T} \zeta_{jt} q_{j(t+(i-1)\Delta T)} \tag{10.2}$$

for $i = 2, \ldots, n$ be the weighted production quantities for item $j = 1, \ldots, J$ after rescheduling in those periods which overlap.[1] The item–specific weights ζ_{jt} for $j = 1, \ldots, J$ and $t = 1, \ldots, T - \Delta T$ should be positive and non–increasing over time to take into account that changing the schedule in periods close to the planning horizon is not as bad as changing the schedule in periods close ahead. For example, one may use

$$\zeta_{jt} \stackrel{def}{=} \frac{1}{t} \tag{10.3}$$

[1] Note, the variables q_{jt} are defined for $t = 1, \ldots, T$ in preceding chapters. It should be clear that when solving the i–th subproblem we have period indices $t = (i-1)\Delta T+1, \ldots, (i-1)\Delta T+T$ and that without loss of generality a simple index transformation enables us to use the presented models and methods as they are.

for $j = 1, \ldots, J$ and $t = 1, \ldots, T - \Delta T$ which is item–independent and thus expresses that we have no preference to keep the schedule for some items more stable than the schedule for some others. Changes of a production plan due to rescheduling should be considered in relation to the total quantity that is scheduled. Hence, a stability measure for generating the production plan of item $j = 1, \ldots, J$ can be defined as the maximum instability

$$sm_j^{max} \stackrel{def}{=} \max \left\{ \frac{| \ddot{q}_j^{(i)} - \dot{q}_j^{(i-1)} |}{\max\{\ddot{q}_j^{(i)}, 1\}} \mid i = 2, \ldots, n \right\} \tag{10.4}$$

or the mean stability

$$sm_j^{mean} \stackrel{def}{=} \frac{1}{n-1} \sum_{i=2}^{n} \frac{| \ddot{q}_j^{(i)} - \dot{q}_j^{(i-1)} |}{\max\{\ddot{q}_j^{(i)}, 1\}}. \tag{10.5}$$

Values close to zero indicate a high stability in both cases. Since we face a multi–item problem, we wish to have a stability measure SM for the generation of an overall plan. Some possibilities are the maximum of the maximum instabilities

$$SM_{max}^{max} \stackrel{def}{=} \max\{sm_j^{max} \mid j = 1, \ldots, J\}, \tag{10.6}$$

the mean of the maximum instabilities

$$SM_{mean}^{max} \stackrel{def}{=} \frac{1}{J} \sum_{j=1}^{J} sm_j^{max}, \tag{10.7}$$

the maximum of the mean stabilities

$$SM_{max}^{mean} \stackrel{def}{=} \max\{sm_j^{mean} \mid j = 1, \ldots, J\}, \tag{10.8}$$

or the mean of the mean stabilities

$$SM_{mean}^{mean} \stackrel{def}{=} \frac{1}{J} \sum_{j=1}^{J} sm_j^{mean}. \tag{10.9}$$

In all cases, a value close to zero indicates good performance. A relation between these four performance measures is established by the following two inequalities:

$$SM_{mean}^{mean} \leq SM_{max}^{mean} \leq SM_{max}^{max} \tag{10.10}$$

$$SM_{mean}^{mean} \leq SM_{mean}^{max} \leq SM_{max}^{max} \tag{10.11}$$

Note, these performance measures are instability measures. Hence, if we would choose $\Delta T = T$, we would have no instability at all. There seems to be a trade–off between the stability of a plan and its cost performance in an ex post analysis. Hence, a cost oriented measure which compares the total costs of the implemented plan with (a lower bound) of the costs of an optimum overall plan that is generated under the assumption that all data for the periods $1, \ldots, \hat{T}$ are known in advance should be considered, too. Choosing a value ΔT that is used for planning is thus a bi–criteria optimization problem.[2]

10.3 Pitfalls: Initial Inventory

When implementing the schedule for the first $\Delta T < T$ periods of a production plan and restarting a lot sizing and scheduling procedure to find a plan for the next T periods, it may happen that some of the released production meets demand in those periods which are currently under review. Hence, initial inventory does indeed occur in real–world situations which underlines what is said in Subsection 3.3.4: Multi–level lot sizing (and scheduling) procedures must take initial inventory into account.

Let us thus back up the demand shuffle heuristic for which we have assumed no initial inventory in Section 6.7 for the sake of convenience. In the presence of initial inventory not every demand that is represented by a node in the data structure Γ_{DS} causes production, because some demand is met by the initial inventory. Using the insight gained in Subsection 3.3.4, the problem is that we do not know in advance which demand is fulfilled by the initial inventory and hence we cannot eliminate some demand to solve the resulting instance under zero initial inventory assumptions. Note, the construction scheme given in Subsection 6.7.4 is capable of taking initial inventory into account. All that remains to be explained is how the data structure and the data structure manipulation changes if we face initial inventory.

[2]For a literature survey of multiple criteria scheduling we refer to [NaHaHe95].

Let us start with the data structure, because in fact nothing changes here. Remarkable to note, however, is that for some nodes $h \in \mathcal{L}$ we now may have $deadline(h) \leq 0$. If this happens to be, the demand of such a node h must be met by the initial inventory, or there must be another node k where $k = succ(h)$ and $deadline(k) \leq 0$, otherwise. Noteworthy to say, if there are nodes with a non–positive deadline entry in the initial data structure, the production that meets the demand represented by these nodes must already have been released to the shop floor.

For the manipulation of the data structure the capacity check (6.91) is a little too conservative, because the right hand side is an upper bound of the actual capacity demand. Even for the initial data structure, this condition might be false. However, we stay with it and do not allow left shift operations which would violate this restriction.

Chapter 11

Method Selection Guidelines

Given a particular PLSP–instance, a question that arises is which method should be chosen to get a production plan with low objective function value. Running all known methods to select the best result afterwards is annoying, if there are hints available which help to make the decision what method to apply. In Section 11.1 we discuss the basic ideas of how to do the method selection. Section 11.2 refines these ideas.

11.1 Basic Ideas

In an implementation of a decision support system we usually find a single method for each problem that is taken into account. For production planning for instance we would expect to have a state–of–the–art algorithm, say demand shuffle, available to solve the lot sizing and scheduling problem.[1] The seemingly ideal case would be that the implemented lot sizing and scheduling method has a good average performance when compared with other available methods and when applied to a wide variety of test instances.

The shortcoming of this strategy is that although a certain method may be superior on average, for special cases some other methods may outperform the seemingly best method. Or, even

[1] Remember Chapter 1 for a discussion of what kind of algorithms can be found in current MRP II software packages.

worse, it may happen that for every instance there is at least one other method which dominates the method with the best on average performance.

From the standpoint of the system developer the choice of a method to be integrated into the system is a decision making problem with a finite number of alternatives and an infinite number of problem parameter settings to which the methods may be applied to. Assume that the space of parameter settings can be grouped into classes of parameter settings, e.g. with almost the same capacity utilization, such that there is a method for each class which dominates the others. The developer then faces a finite though large number of cases the system must deal with. The decision making process is a process under uncertainty, because of two reasons. First, the customers who will use the decision support system are usually unknown during the developing phase. And second, the parameter settings of a customer, e.g. the capacity utilizations of the machines, vary over run–time. Decision theory does not give a unique answer for this decision making problem, because several decision rules are suggested (see [BaCo92, Bit81, Din82, EiWe93, Fre88, Hol79, Lau91, Lau93, Sal88] for a review) which may yield different results.

What we propose is to give up the idea to have only one method implemented and to maintain a database which provides for several parameter level combinations (see Chapter 4) a method to be applied. This database is filled with entries by performing a computational study with a (full) factorial experimental design. The method that performs best on average for a certain parameter level combination is entered into the database. Later, when being used by the customer, the instance to be solved is attacked by the method that is looked up in the database and that corresponds to the parameter level combination of the instance at hand. In other words, the decision which method to apply should be done taking the instance characteristics into account.

The problem which remains is that there is an infinite number of potential parameter level combinations on the one hand. Consider for instance the setup and holding costs which are real–valued and unbounded and thus, by definition, have an infinite domain.

On the other hand we must confine ourself to a finite number of parameter level combinations when doing a computational study. Hence, we face the problem that later, when a particular instance is to be solved, the parameter values of that instance will usually not match to one of the parameter settings in the database. Loosely speaking, this means that we need some kind of mapping function which finds for each instance to be solved a record in the database with "almost equal" parameter settings.

The problem of finding such a record can be viewed as a pattern recognition problem. The parameter values of the instance to be solved is the pattern to be recognized and the parameter settings in the database are those patterns that are known. For solving pattern recognition problems several approaches are proposed. The traditional ones are based on a model for message transmission [Bla83] where the messages received may differ from the codewords that are sent due to noise and distortion. Hence, the received words are to be recognized. Often, a distance measure which computes the distance of the pattern to be recognized to each of the known patterns is used.[2] As a result a pattern is recognized as the known pattern that gives the smallest distance. Beside this, there are artificially intelligent approaches utilizing neural networks for pattern recognition [CoKu89, Fuk88].

The problem of selecting an appropriate method to solve a given instance can also be seen as a problem amenable to so–called case–based reasoning which is a more and more evolving approach for decision making. The basic principles of case–based reasoning equal those described above, i.e. a database with previous cases, or instances as we would say, and a way to retrieve and to suggest a solution for a new case. In addition, case–base reasoning often incorporates learning which means that the database is augmented during run–time. For more details about case–based reasoning we refer to [AaPl94, AlBa96].

[2] An example of a well–known distance measure is the Hamming distance which counts the number of those positions in two codewords at which different letters are found.

11.2 Instance Characterization

For the PLSP–MM the instances can be characterized according to what is done in Section 4.2. Note, real–world instances are not generated by an instance generator and thus some values of interest are not directly given, but must be derived from the given data instead. In the following we enumerate some values which should be considered: J, the number of items and T, the planning horizon. $\frac{J}{M} \in [1, \ldots, J]$, the average number of items per machine. $\mathcal{C} \in [0, 1]$, the complexity of the gozinto–structure. $\frac{1}{J} \sum_{j=1}^{J} \frac{s_j}{h_j}$, the average ratio of setup and holding costs. $U = \frac{100}{M \cdot T} \sum_{m=1}^{M} \sum_{t=1}^{T} \frac{\hat{C}_{mt}}{C_{mt}}$, the average capacity utilization where $\hat{C}_{mt}$ is the capacity the instance generator APCIG would have generated for a 100% capacity utilization. $\frac{1}{J} \sum_{j=1}^{J} \frac{nr_j}{T}$, the average size of the lots within a period.

For getting points of references, instances with certain characteristics can be generated with the instance generator given in Chapter 4. Remarkably, these instances need not to be solved optimally, because we are interested in the best available method for a given parameter level combination. Hence, it suffices to compare the heuristic results which allows us to have rather large instances to be included.

As an example, let us assume that we have a database available which helps to choose a method where the gozinto–structure complexities in $[0, \ldots, 0.5[$ and in $[0.5, \ldots, 1]$ are discriminated, and the setup and holding cost ratios are grouped with respect to orders of magnitude, i.e. $\log_{10}\left(\frac{1}{J} \sum_{j=1}^{J} \frac{s_j}{h_j}\right)$ is in $]-\infty, \ldots, 1.5[$, $[1.5, \ldots, 2.5[$, $[2.5, \ldots, 3.5[$, or $[3.5, \ldots, \infty[$, respectively. Furthermore, assume that for the instance to be solved $\mathcal{C} = 0.8$ and $\frac{1}{J} \sum_{j=1}^{J} \frac{s_j}{h_j} = 1,000$ holds. The method to choose can then be retrieved from the database by accessing the record for a complexity out of $[0.5, \ldots, 1]$, a logarithmical cost ratio out of $[2.5, \ldots, 3.5[$, and all other parameter values as determined by the instance to be solved.

Chapter 12

Research Opportunities

In this chapter we give some hints for future work. Clearly, one should pursuit to find more methods for the presented models in order to improve the current bounds. Beside this, effort should be spent to extend the models and the methods for further real–world requirements. Section 12.1 summarizes some of these extensions. To be integrated into a real–world environment with harmony, one must take into account that lot sizing and scheduling interacts with other planning tasks. Hence, Section 12.2 reveals some links to other planning problems.

12.1 Extensions

Time–varying costs: Instead of using cost parameters that are constant over time, using time–varying costs would be more general. If so, production quantity dependent costs should also be taken into account. A motivation for having such a general cost structure would be its use as a control instrument for the generation of production plans. Periods in which production should only take place if no feasible plan can be found otherwise should have high setup and production costs. Such periods could for instance be considered as overtime periods. The single–item CLSP with time–variant costs is dealt with in [ChFlLi94]. [Hin96] attacks the multi–item, single–level CLSP. The single–level DLSP with a time–variant cost structure is presented in [HoKo94]. A review of single–item lot sizing is provided in [Wol95].

Joint and item–specific setup costs: In the CLSP setup costs are charged for every item for which production takes place in a period. Hence, we face item–specific setup costs. Since at most one setup per period can happen in small bucket models (DLSP, CSLP, and PLSP) we face item–specific rather than joint setup costs, too. Large bucket models which combine joint and item–specific setup costs are referred to as joint replenishment problem models [AkEr88, ChMe94, Ere88, FeZh92, FeTz94a, JaMaMu85, Jon90, Kao79, Kir95]. For small bucket models with multiple machines several items may be scheduled within a period, too. Hence, considering joint setup costs may be of relevance.

Setup times: Setting a machine up does not only cause setup costs, but requires some of the available (time) capacity, too. Hence, it is of great practical importance to take setup times into account. For the CLSP, models and methods with setup times are discussed in [BiMClTh86, BlMi84, CaMaWa90, Der95, DiBaKaZi92a, DiBaKaZi92b, DzGo65, Hel94, LoLaOn91, TeDe93, TeDe95, TrThMCl89]. The DLSP with setup times is concerned about in [CaSaKuWa93]. Models for the PLSP with setup times are given in [DrHa95, Haa94].

Sequence dependencies: In many real–world applications, the setup costs and times are sequence dependent. Hence, it seems to be important to have models and methods which handle sequence dependencies as well. By definition, sequence dependencies can only be taken into account if we solve a lot sizing and scheduling problem. Hence, the CLSP approach is not appropriate for such cases. The DLSP with sequence dependent costs is considered in [Fle94, Pop93, Sch82]. In [SaSoWaDuDa95] the focus is on the DLSP with sequence dependent setup costs and times. A PLSP–model with sequence dependent setup costs is given in [Haa94]. [FlMe96, Haa96] tackle large bucket lot sizing and scheduling problems with sequence dependent setup costs. A large bucket problem with sequence dependent setup costs and times is solved in [HaKi96]. An uncapacitated lot sizing problem with sequence dependent setup costs is considered in [DiRa89]. [Dob92] attacks a lot sizing and scheduling problem with sequence dependent setup costs and times where demand is stationary.

Learning effects in setups and production: Some authors investigate the effects of setup cost and time as well as production time reduction due to learning. These reductions may occur at the very beginning of a production process. Hence, it is quite controversial if learning effects are to be considered for short–term lot sizing, especially on a rolling horizon basis. A Wagner–Whitin type of model with setup learning is discussed in [MaWa93]. [ChSe90, PrCaRa94] study the CLSP with setup learning. The ELSP with learning effects is treated in [Cha89, KaMaMo88, SmMo85] and a problem under EOQ–assumptions is dealt with in [MuSp83].

Backlogs and stockouts: In practice, it will often be the case that the due dates of the orders are not strictly enforced. Hence, backlogging may occur to avoid overtime. However, backlogging bears the risk that orders are withdrawn (stockouts) and that the customer's good will is lost. Thus, backlogging incurs high penalty costs. Some early work is reported in [HvL70a, Zan69]. Heuristics for Wagner–Whitin–models with backlogging are presented in [ChCh90, BlKu74, FeTz93, GuBr92, HsLaCh92, Web89]. A PLSP–model with backordering and stockouts is given in [Haa94].

Material supply constraints: Scarce capacities in the supply chain may impose certain restrictions on the availability of raw material. In [LeGaXi93] the DLSP is extended for such constraints.

Safety stocks: One way to hedge demand uncertainties is to maintain item–specific safety stocks. For the single–level DLSP, safety stocks are considered in [Fle90].[1]

Capacitated storage space: The (weighted) sum of items that are held in inventory at the end of a period may be restricted by capacity constraints. A problem with such kind of constraints is referred to as the warehouse scheduling problem (WSP) [Hal88, HaJa95, PaPa76, Zol77] which is closely related to lot sizing.

[1]Note, for multi–level gozinto–structures the problem of determining safety stocks is more complex than in the single–level case, because instead of maintaining safety stocks for items with external demand we may decide to have safety stocks for component parts [Ind95].

12.2 Linkages to Other Areas

Make–or–buy decisions: Though outsourcing typically is a long-term decision [Gil88, Hom95, Män81, PiFr93], make–or–buy decisions for the short–term provide a means to find feasible production plans if capacity is temporarily exhausted or to reduce production costs by getting bargains for component parts. A CLSP–like model with make–or–buy decisions is presented in [LeZi89]. A single–level PLSP–model is given in [Haa94]. For the multi–level CLSP and the multi–level PLSP, models and methods are outlined in [Kim95].

Substitution between products: For some end items there is a degree of freedom which component parts are to be assembled. This is to say that some parts may substitute each other either in full or in partial. The decision which part type is used usually affects the gozinto–structure to be considered. Also, there may exist a resource–time trade–off, because some parts may require less capacity to be produced, but incur higher costs when manufactured. For an EOQ–model, the effects of substitutions between products is investigated in [DrGuPa95].

Distribution planning: Distributing finished goods to customers interacts with the production process. On the one hand, holding costs are incurred if end items are produced but not delivered immediately. On the other hand, not every end item that is finished can be distributed instantly when truck capacities are scarce. Moreover, the tour of a vehicle is determined by the items that are loaded to be delivered. This coordination problem is dealt with in [ChFi94].

Cutting and packing:[2] Before items can be manufactured some input parts are to be cut. After end items are finished they need to be packed. In both cases, the decision what items to cut or to pack, respectively, usually depends on the production plan for an interval of periods much smaller than the planning horizon for lot sizing (and scheduling). Thus, lot sizing affects the waste of raw materials during the cutting process and the waste of space during packing. On the other hand, planning for cutting imposes material

[2] For an introduction into cutting and packing we refer to [Dyc90].

supply constraints and thus has an impact on lot sizing, too.

Project scheduling: Scheduling a project may generate (external) demand for items that are to be produced in advance. The deadline for this demand depends on the project's schedule. From another point of view, the production plan defines the number of items that are available in a certain period and thus determines some of the resource limits that are used as an input for project scheduling. Hence, there is an interaction between project scheduling and lot sizing, too.

Chapter 13

Conclusion

This book deals with multi–level lot sizing and scheduling and most chapters are devoted to present methods for capacitated, dynamic and deterministic models.

In **Chapter 1** we outline the problem of lot sizing and scheduling as to be done for short–term production planning in many firms. A key issue for the motivation of this work is that the traditional MRP II logic, which is implemented in most modern software packages that are commercially distributed, fails to solve the multi–level lot sizing and scheduling problem. It can be shown that high work in process, long lead times, and excessive need for overtime about which practitioners complain are often not as fate would have it, but a consequence of strictly following the MRP II concept.

To be sure that this work is indeed original and new, **Chapter 2** contains the most comprehensive literature review of multi–level lot sizing that is available today. It turns out that many real–world requirements are not treated with. Many authors do not even consider capacity constraints. Those who do make many restrictive assumptions. For example, they confine their work to a single bottleneck resource or to a certain type of gozinto–structure. The multi–level capacitated lot sizing problem with multiple machines has attracted only few researchers in the recent past. Nevertheless, these few people have gained remarkable results. For lot sizing and scheduling almost all researchers focus on the single–level case.

Only very few attempts are made to attack multi–level lot sizing and scheduling. In summary, the problem of simultaneous lot sizing and scheduling with multi–level gozinto–structures turns out to be completely unresolved. The fact that this important problem has not been tackled indicates that it is a real challenge to do so.

In **Chapter 3** we define the assumptions under which this work is done. A MIP–model formulation of the so–called multi–level PLSP–MM is presented to give a precise definition of what the problem is. It is shown that this model is a generalization of other well–known problems, namely the DLSP and the CSLP. Hence, the subtitle of this book makes sense, because any method for the PLSP will also be able to solve the DLSP or the CSLP. The features of the model that are most remarkable are general gozinto–structures and multiple machines. But other details such as time–varying capacities or general cost structures are also worth to be mentioned. Several extensions are modelled and discussed, too. These are parallel machines, multiple resources, and partially renewable resources. All these models represent cases which are highly relevant for practical purposes, but have not been treated in the literature even not under single–level assumptions (there are a few exceptions for parallel machines). Other model extensions which are more or less standard (e.g. sequence dependencies, or setup times) are not given which does not mean that they are easy to solve. We also derive some first insight into the multi–level PLSP and show that initial inventory cannot be neglected without loss of generality as most authors do. Finally, we develop a postprocessor which helps to improve feasible production plans.

To evaluate methods for multi–level lot sizing and scheduling one should use a carefully designed collection of test instances. Hence, **Chapter 4** defines a parameter controlled instance generator which allows to generate test–beds systematically. Moreover, we specify a set of 1,080 small PLSP–MM–instances which can be solved optimally with standard solvers. The definition of the instance generator is an important contribution, because up to now there is neither an established standard test–bed nor an instance generator available from the literature.

In **Chapter 5** we discuss lower bounds for the PLSP–MM. It turns out that the LP–relaxation of the original model formulation as well as the LP–relaxation of a simple plant location reformulation gives only poor results. This is interesting to note, because a plant location model for lot sizing without scheduling is proposed in the literature because of its remarkable results. A tailor–made branch–and–bound procedure is developed to solve the uncapacitated PLSP–MM. The results obtained are decidedly better than the results of the LP–relaxation. Further improvements are gained via a Lagrangean relaxation of the capacity constraints. Within one hour, lower bounds with an average deviation of -5.59% from the optimum results are computed. However, due to its run–time performance it will only be of use for small instances. More efficient procedures are left for future work.

Chapter 6 presents five classes of tailor–made heuristics for the PLSP–MM. A precise specification is given for a randomized regret based sampling procedure, a cellular automaton, a genetic algorithm, a disjunctive arc based tabu search, and a so–called demand shuffle method. The common idea of all these approaches is to construct production plans in a backward oriented manner. This allows us to take multi–level gozinto–structures and initial inventories neatly into account and to perform efficient feasibility checks. An in–depth computational study with small instances reveals the dominance of the demand shuffle procedure with respect to both, the average deviation from the optimum result and the infeasibility ratio. The average results are a 6.60% deviation and an amazing 0.58% infeasibility ratio which are computed within 0.44 CPU–seconds. A comparison with the results of the other methods shows that these are remarkable figures which prove that the demand shuffle heuristic makes a significant contribution. For large instances, however, randomized regret based sampling turns out to be a challenging competitor as long as the number of items is low, say less than 10. For more items, demand shuffle again outperforms the other methods. A benchmark for future work has certainly been set. The findings for the cellular automaton, the genetic algorithm, and the disjunctive arc based tabu search are disappointing. Particularly, the disjunctive arc based procedure yields outrageously bad results for large instances, because it rarely finds feasible solutions. Noteworthy to say that with respect to the run–time performance the genetic algorithm cannot be beaten. This gives the opportunity to evaluate many individuals which, when done, improves the results of the genetic algorithm.

A method for lot sizing and scheduling with parallel machines is presented in **Chapter 7**. **Chapter 8** describes a heuristic for the multiple resource case, and **Chapter 9** introduces a solution procedure for partially renewable resources. In all cases the proposed methods are modifications of the demand shuffle heuristic. Preliminary computational studies with very small test–beds are done. The results are compared with those for LP–relaxations, because optimum solutions cannot be computed within reasonable time, say within less than three hours. Unfortunately, the gaps between upper and lower bounds are dramatically high. A definite statement about the performance of the heuristics can therefore not been made. Other methods should be developed in the future and be compared with the demand shuffle variants.

Since lot sizing and scheduling will in practice be done on the basis of a rolling horizon, **Chapter 10** discusses the problems related to such implementations. A key issue is the fact that initial inventory must be taken into account and cannot be neglected as many authors do. Furthermore, we suggest performance measures based on the nervousness of production plans due to rescheduling.

An instance–specific choice of (lot sizing and scheduling) methods is proposed in **Chapter 11**. The problem of doing so is identified as being a pattern recognition problem and closely relates to case–based reasoning. Moreover, we discuss some instance characteristics for lot sizing and scheduling instances which can be derived from the parameters of the instance generator presented in earlier chapters.

Some ideas for future work are compiled in **Chapter 12**. Among them are some straightforward extensions such as sequence dependencies and setup times. But linkages to other areas with which interaction effects exist are also revealed. A challenging task for future research will be to integrate lot sizing and scheduling into real–world situations.

Appendix A

Lower Bounds for the PLSP–MM

This appendix contains the details of the computational results which are summarized in Chapter 5. Four ways to obtain lower bounds are tested: Solving the LP–relaxation of the PLSP–MM–model, the LP–relaxation of a simple plant location model, the PLSP–MM with relaxed capacity constraints, and a Lagrangean relaxation of the capacity constraints.

The test–bed as described in Chapter 4 consists of 1,080 small PLSP–MM–instances, 1,033 of them have a feasible solution. It is designed using a full factorial design where the following parameters are systematically varied: The capacity utilization U, the number of machines M, the complexity of the gozinto–structures $\mathcal{C}$, the ratio of setup and holding costs $COSTRATIO$, and the demand pattern $(T_{macro}, T_{micro}, T_{idle})$. For each parameter level combination we have (at most) 10 instances with a feasible solution.

The following tables show three numbers for each parameter level combination: The average deviation from the optimum objective function value, the worst case deviation from the optimum objective function value, and the number of instances that are considered.

A.1 LP–Relaxation: PLSP–Model

			$COSTRATIO = 5$		
U	M	C	(10,1,5)	(5,2,2)	(1,10,0)
30	1	0.2	-41.91 [-55.09/10]	-39.94 [-52.31/10]	-19.02 [-28.12/10]
30	1	0.8	-42.88 [-52.14/10]	-41.91 [-50.87/10]	-22.26 [-26.94/10]
30	2	0.2	-18.39 [-33.86/10]	-17.84 [-28.75/10]	-8.42 [-17.89/10]
30	2	0.8	-18.39 [-28.33/9]	-18.19 [-29.45/10]	-9.14 [-19.04/10]
50	1	0.2	-44.41 [-56.49/10]	-41.87 [-52.57/10]	-13.81 [-22.09/10]
50	1	0.8	-45.34 [-53.65/10]	-44.67 [-53.30/10]	-19.57 [-25.09/10]
50	2	0.2	-21.96 [-34.34/10]	-19.61 [-26.54/10]	-9.30 [-17.20/10]
50	2	0.8	-19.57 [-32.06/10]	-17.16 [-29.55/10]	-10.63 [-17.18/10]
70	1	0.2	-37.93 [-45.21/10]	-40.91 [-47.69/8]	-13.16 [-22.15/10]
70	1	0.8	-48.34 [-61.47/8]	-44.98 [-50.07/6]	-19.81 [-26.37/10]
70	2	0.2	-16.53 [-25.20/10]	-17.82 [-28.27/7]	-10.12 [-18.84/10]
70	2	0.8	-16.66 [-32.83/9]	-16.12 [-25.61/8]	-12.16 [-20.00/10]

Table A.1: LP–Relaxation of the PLSP–Model, Part 1

			$COSTRATIO = 150$		
U	M	C	(10,1,5)	(5,2,2)	(1,10,0)
30	1	0.2	-57.71 [-72.97/10]	-62.92 [-76.63/10]	-61.25 [-70.39/10]
30	1	0.8	-52.90 [-64.31/10]	-56.27 [-66.78/10]	-53.54 [-63.51/10]
30	2	0.2	-46.12 [-56.29/10]	-51.29 [-60.50/10]	-46.48 [-53.36/10]
30	2	0.8	-37.25 [-49.90/10]	-44.27 [-52.48/9]	-41.09 [-49.96/10]
50	1	0.2	-59.48 [-72.57/10]	-63.23 [-75.70/10]	-55.61 [-65.18/10]
50	1	0.8	-54.29 [-65.62/10]	-57.75 [-68.05/10]	-51.76 [-56.27/10]
50	2	0.2	-48.48 [-58.35/10]	-53.22 [-61.28/10]	-44.51 [-50.14/10]
50	2	0.8	-39.38 [-53.04/10]	-44.24 [-53.50/10]	-42.63 [-47.98/10]
70	1	0.2	-55.06 [-66.56/10]	-63.05 [-71.87/8]	-52.60 [-60.53/10]
70	1	0.8	-55.66 [-63.86/8]	-59.59 [-66.27/6]	-51.82 [-55.10/9]
70	2	0.2	-43.83 [-49.81/10]	-53.57 [-59.46/7]	-44.91 [-52.25/10]
70	2	0.8	-37.48 [-46.61/10]	-44.68 [-50.40/8]	-44.23 [-49.22/10]

Table A.2: LP–Relaxation of the PLSP–Model, Part 2

			$COSTRATIO = 900$		
U	M	C	(10,1,5)	(5,2,2)	(1,10,0)
30	1	0.2	-74.11 [-83.45/10]	-78.17 [-85.27/10]	-82.27 [-85.71/10]
30	1	0.8	-66.34 [-76.74/10]	-71.41 [-80.47/10]	-78.35 [-85.07/10]
30	2	0.2	-60.52 [-71.60/10]	-64.12 [-72.45/10]	-64.44 [-71.52/10]
30	2	0.8	-54.04 [-64.24/9]	-59.80 [-68.88/9]	-62.39 [-71.45/10]
50	1	0.2	-73.68 [-82.24/10]	-77.20 [-83.04/10]	-80.11 [-83.38/10]
50	1	0.8	-67.39 [-76.35/10]	-72.96 [-79.49/10]	-77.32 [-82.69/10]
50	2	0.2	-62.76 [-69.97/10]	-66.95 [-73.18/10]	-63.14 [-69.17/10]
50	2	0.8	-58.19 [-65.62/10]	-64.88 [-72.02/9]	-62.91 [-69.74/9]
70	1	0.2	-74.92 [-81.19/10]	-81.53 [-83.93/8]	-78.45 [-81.29/10]
70	1	0.8	-71.71 [-76.17/8]	-78.31 [-83.05/6]	-76.72 [-80.47/10]
70	2	0.2	-63.47 [-68.97/10]	-71.84 [-75.28/7]	-64.03 [-67.71/10]
70	2	0.8	-59.86 [-69.34/10]	-69.83 [-76.17/8]	-64.02 [-68.79/10]

Table A.3: LP–Relaxation of the PLSP–Model, Part 3

A.2 LP–Relaxation: Plant Location Model

			$COSTRATIO = 5$		
U	M	C	(10,1,5)	(5,2,2)	(1,10,0)
30	1	0.2	-22.79 [-31.11/10]	-20.96 [-29.68/10]	-14.06 [-16.89/10]
30	1	0.8	-32.76 [-35.47/10]	-31.83 [-35.11/10]	-19.11 [-23.75/10]
30	2	0.2	-8.37 [-17.43/10]	-10.18 [-15.50/10]	-7.87 [-17.83/10]
30	2	0.8	-11.67 [-21.86/9]	-12.82 [-23.63/10]	-8.70 [-19.02/10]
50	1	0.2	-26.15 [-33.78/10]	-25.49 [-34.51/10]	-12.52 [-21.13/10]
50	1	0.8	-36.33 [-39.87/10]	-36.13 [-38.84/10]	-18.96 [-24.49/10]
50	2	0.2	-12.63 [-18.86/10]	-14.02 [-17.02/10]	-9.23 [-17.29/10]
50	2	0.8	-13.84 [-27.18/10]	-13.67 [-26.34/10]	-10.52 [-16.98/10]
70	1	0.2	-26.80 [-35.95/10]	-29.05 [-35.78/8]	-12.79 [-21.33/10]
70	1	0.8	-43.26 [-59.85/8]	-39.04 [-44.17/6]	-19.57 [-26.29/10]
70	2	0.2	-14.80 [-21.94/10]	-16.82 [-27.23/7]	-10.26 [-19.26/10]
70	2	0.8	-15.67 [-30.82/9]	-15.16 [-24.78/8]	-12.18 [-19.98/10]

Table A.4: LP–Relaxation of the Plant Location Model, Part 1

			$COSTRATIO = 150$		
U	M	C	(10,1,5)	(5,2,2)	(1,10,0)
30	1	0.2	-41.60 [-49.75/10]	-48.65 [-56.25/10]	-53.06 [-59.06/10]
30	1	0.8	-43.42 [-50.24/10]	-46.84 [-53.27/10]	-47.52 [-53.46/10]
30	2	0.2	-37.00 [-43.04/10]	-43.46 [-50.15/10]	-43.37 [-50.30/10]
30	2	0.8	-30.58 [-42.46/10]	-38.18 [-47.69/9]	-38.09 [-46.67/10]
50	1	0.2	-44.88 [-50.62/10]	-51.71 [-56.75/10]	-51.19 [-59.85/10]
50	1	0.8	-46.15 [-52.91/10]	-50.17 [-56.48/10]	-48.25 [-52.24/10]
50	2	0.2	-40.82 [-48.67/10]	-47.97 [-54.33/10]	-44.16 [-50.40/10]
50	2	0.8	-33.87 [-46.85/10]	-40.35 [-49.96/10]	-41.48 [-49.29/10]
70	1	0.2	-45.19 [-51.61/10]	-53.92 [-58.86/8]	-50.65 [-59.46/10]
70	1	0.8	-50.27 [-61.97/8]	-53.92 [-58.59/6]	-49.81 [-52.98/9]
70	2	0.2	-40.64 [-45.89/10]	-50.70 [-57.03/7]	-45.82 [-53.64/10]
70	2	0.8	-35.41 [-45.24/10]	-42.86 [-46.68/8]	-43.96 [-50.96/10]

Table A.5: LP–Relaxation of the Plant Location Model, Part 2

U	M	C	COSTRATIO = 900 (10,1,5)	(5,2,2)	(1,10,0)
30	1	0.2	-61.77 [-68.58/10]	-67.61 [-72.17/10]	-76.76 [-80.35/10]
30	1	0.8	-57.20 [-64.08/10]	-62.70 [-68.87/10]	-72.87 [-77.78/10]
30	2	0.2	-51.84 [-58.42/10]	-57.27 [-62.98/10]	-61.88 [-70.11/10]
30	2	0.8	-47.40 [-54.24/9]	-53.35 [-63.11/9]	-59.17 [-65.21/10]
50	1	0.2	-63.24 [-69.66/10]	-68.64 [-72.90/10]	-77.09 [-80.53/10]
50	1	0.8	-60.20 [-65.80/10]	-66.31 [-70.00/10]	-74.09 [-78.12/10]
50	2	0.2	-56.36 [-63.81/10]	-62.41 [-70.90/10]	-62.72 [-70.46/10]
50	2	0.8	-53.00 [-61.13/10]	-60.59 [-65.52/9]	-61.63 [-66.29/9]
70	1	0.2	-68.93 [-72.32/10]	-76.09 [-77.67/8]	-77.20 [-80.64/10]
70	1	0.8	-67.76 [-70.38/8]	-74.27 [-78.33/6]	-75.09 [-78.27/10]
70	2	0.2	-60.13 [-67.80/10]	-69.14 [-73.71/7]	-64.77 [-69.86/10]
70	2	0.8	-57.42 [-67.40/10]	-67.30 [-71.99/8]	-63.71 [-67.36/10]

Table A.6: LP–Relaxation of the Plant Location Model, Part 3

A.3 Capacity Relaxation

			$COSTRATIO = 5$		
U	M	C	(10,1,5)	(5,2,2)	(1,10,0)
30	1	0.2	-0.72 [-4.04/10]	-0.78 [-3.28/10]	-5.55 [-7.67/10]
30	1	0.8	-2.43 [-4.18/10]	-3.32 [-8.62/10]	-4.97 [-10.88/10]
30	2	0.2	-1.10 [-5.30/10]	-3.42 [-8.56/10]	-19.29 [-41.28/10]
30	2	0.8	-0.56 [-2.32/9]	-2.58 [-11.96/10]	-17.42 [-35.26/10]
50	1	0.2	-5.09 [-10.23/10]	-6.66 [-13.56/10]	-19.64 [-26.23/10]
50	1	0.8	-7.61 [-12.49/10]	-9.38 [-18.32/10]	-14.57 [-22.24/10]
50	2	0.2	-5.87 [-8.53/10]	-9.81 [-15.04/10]	-40.88 [-55.36/10]
50	2	0.8	-3.88 [-9.14/10]	-5.37 [-16.61/10]	-29.10 [-48.28/10]
70	1	0.2	-9.59 [-19.56/10]	-13.70 [-18.97/8]	-33.60 [-46.74/10]
70	1	0.8	-18.58 [-41.16/8]	-12.46 [-21.52/6]	-24.10 [-35.12/10]
70	2	0.2	-20.33 [-30.13/10]	-20.54 [-23.81/7]	-51.54 [-63.15/10]
70	2	0.8	-10.62 [-19.07/9]	-11.11 [-15.43/8]	-37.57 [-56.92/10]

Table A.7: Capacity Relaxation, Part 1

U	M	C	$COSTRATIO = 150$ (10,1,5)	(5,2,2)	(1,10,0)
30	1	0.2	-1.13 [-3.35/10]	-2.70 [-7.39/10]	-3.22 [-5.43/10]
30	1	0.8	-3.86 [-6.78/10]	-4.61 [-9.17/10]	-5.45 [-10.85/10]
30	2	0.2	-3.37 [-7.49/10]	-7.79 [-12.54/10]	-15.42 [-70.81/10]
30	2	0.8	-1.47 [-5.92/10]	-5.93 [-8.74/9]	-11.51 [-19.37/10]
50	1	0.2	-6.54 [-16.46/10]	-8.32 [-16.40/10]	-8.92 [-15.34/10]
50	1	0.8	-8.50 [-13.35/10]	-10.55 [-18.49/10]	-13.78 [-22.57/10]
50	2	0.2	-9.24 [-14.88/10]	-15.89 [-24.27/10]	-19.79 [-32.50/10]
50	2	0.8	-6.19 [-13.27/10]	-11.42 [-16.09/10]	-23.40 [-32.08/10]
70	1	0.2	-11.56 [-23.18/10]	-13.14 [-24.55/8]	-17.12 [-28.73/10]
70	1	0.8	-16.32 [-37.62/8]	-14.57 [-20.44/6]	-22.85 [-30.05/9]
70	2	0.2	-19.07 [-31.67/10]	-23.51 [-34.64/7]	-29.98 [-46.68/10]
70	2	0.8	-13.45 [-24.79/10]	-18.21 [-22.14/8]	-32.37 [-42.12/10]

Table A.8: Capacity Relaxation, Part 2

			$COSTRATIO = 900$		
U	M	C	(10,1,5)	(5,2,2)	(1,10,0)
30	1	0.2	-1.13 [-2.43/10]	-1.96 [-4.32/10]	-0.79 [-1.36/10]
30	1	0.8	-3.91 [-7.27/10]	-8.56 [-48.45/10]	-2.17 [-3.94/10]
30	2	0.2	-16.40 [-68.30/10]	-5.62 [-23.28/10]	-2.76 [-11.11/10]
30	2	0.8	-6.79 [-14.58/9]	-7.54 [-20.78/9]	-4.08 [-7.53/10]
50	1	0.2	-4.85 [-8.56/10]	-4.98 [-11.79/10]	-2.34 [-4.73/10]
50	1	0.8	-10.47 [-20.27/10]	-12.69 [-30.61/10]	-6.37 [-12.29/10]
50	2	0.2	-12.33 [-25.98/10]	-17.27 [-33.69/10]	-5.66 [-13.90/10]
50	2	0.8	-16.50 [-27.72/10]	-19.57 [-29.76/9]	-9.50 [-14.73/9]
70	1	0.2	-22.29 [-29.83/10]	-26.38 [-35.87/8]	-4.99 [-9.94/10]
70	1	0.8	-27.04 [-39.88/8]	-29.66 [-35.32/6]	-11.27 [-20.12/10]
70	2	0.2	-24.81 [-42.10/10]	-29.26 [-42.48/7]	-11.96 [-34.45/10]
70	2	0.8	-27.27 [-39.53/10]	-35.33 [-43.75/8]	-16.20 [-28.71/10]

Table A.9: Capacity Relaxation, Part 3

A.4 Lagrangean Relaxation

			$COSTRATIO = 5$		
U	M	C	(10,1,5)	(5,2,2)	(1,10,0)
30	1	0.2	-0.49 [-2.38/10]	-0.44 [-3.28/10]	-2.64 [-7.01/10]
30	1	0.8	-1.98 [-4.07/10]	-2.06 [-6.14/10]	-2.90 [-6.74/10]
30	2	0.2	-0.62 [-3.18/10]	-0.80 [-3.34/10]	-5.63 [-19.87/10]
30	2	0.8	-0.32 [-1.49/9]	-1.33 [-7.43/10]	-3.37 [-16.14/10]
50	1	0.2	-3.22 [-5.67/10]	-3.77 [-6.13/10]	-5.46 [-7.83/10]
50	1	0.8	-2.25 [-6.84/10]	-2.86 [-6.20/10]	-6.66 [-13.56/10]
50	2	0.2	-4.00 [-6.19/10]	-5.52 [-8.80/10]	-7.14 [-19.68/10]
50	2	0.8	-2.53 [-7.29/10]	-2.77 [-9.83/10]	-6.27 [-19.52/10]
70	1	0.2	-3.87 [-12.12/10]	-7.93 [-16.84/8]	-5.04 [-10.59/10]
70	1	0.8	-9.89 [-27.30/8]	-5.86 [-9.33/6]	-9.44 [-19.41/10]
70	2	0.2	-12.50 [-22.89/10]	-11.39 [-13.84/7]	-7.68 [-22.12/10]
70	2	0.8	-7.47 [-17.17/9]	-6.59 [-12.10/8]	-6.78 [-18.16/10]

Table A.10: Lagrangean Relaxation, Part 1

			$COSTRATIO = 150$		
U	M	C	(10,1,5)	(5,2,2)	(1,10,0)
30	1	0.2	-0.71 [-2.74/10]	-1.44 [-4.26/10]	-1.83 [-3.61/10]
30	1	0.8	-2.32 [-4.39/10]	-1.48 [-3.50/10]	-1.86 [-4.23/10]
30	2	0.2	-4.54 [-29.54/10]	-2.43 [-6.10/10]	-2.91 [-11.03/10]
30	2	0.8	-1.13 [-5.91/10]	-2.81 [-4.54/9]	-5.46 [-13.31/10]
50	1	0.2	-3.69 [-11.13/10]	-2.86 [-6.51/10]	-2.26 [-5.45/10]
50	1	0.8	-1.90 [-4.17/10]	-1.77 [-4.18/10]	-4.92 [-13.78/10]
50	2	0.2	-5.57 [-11.58/10]	-7.51 [-13.00/10]	-6.35 [-18.43/10]
50	2	0.8	-4.79 [-11.97/10]	-5.55 [-11.62/10]	-10.20 [-18.02/10]
70	1	0.2	-3.52 [-9.59/10]	-4.16 [-7.42/8]	-3.68 [-11.21/10]
70	1	0.8	-7.04 [-21.67/8]	-3.84 [-6.98/6]	-8.66 [-16.04/9]
70	2	0.2	-12.94 [-25.87/10]	-11.16 [-22.50/7]	-11.27 [-30.24/10]
70	2	0.8	-11.29 [-22.48/10]	-11.47 [-19.36/8]	-14.10 [-23.50/10]

Table A.11: Lagrangean Relaxation, Part 2

U	M	C	$COSTRATIO = 900$ (10,1,5)	(5,2,2)	(1,10,0)
30	1	0.2	-0.55 [-1.77/10]	-7.56 [-65.22/10]	-0.31 [-0.84/10]
30	1	0.8	-1.25 [-2.33/10]	-1.22 [-3.48/10]	-0.75 [-1.73/10]
30	2	0.2	-1.28 [-6.37/10]	-3.31 [-19.30/10]	-1.57 [-10.17/10]
30	2	0.8	-3.64 [-7.95/9]	-4.67 [-19.68/9]	-2.01 [-5.44/10]
50	1	0.2	-1.31 [-5.68/10]	-0.96 [-4.80/10]	-0.58 [-2.23/10]
50	1	0.8	-2.30 [-5.43/10]	-3.15 [-10.27/10]	-2.76 [-7.60/10]
50	2	0.2	-6.92 [-20.63/10]	-12.01 [-27.65/10]	-2.70 [-12.16/10]
50	2	0.8	-11.03 [-24.15/10]	-14.62 [-25.89/9]	-6.17 [-12.33/9]
70	1	0.2	-10.32 [-22.21/10]	-16.05 [-26.99/8]	-1.39 [-4.41/10]
70	1	0.8	-12.88 [-21.16/8]	-16.84 [-25.95/6]	-5.06 [-12.22/10]
70	2	0.2	-20.37 [-40.73/10]	-24.17 [-37.70/7]	-7.62 [-32.68/10]
70	2	0.8	-22.94 [-39.53/10]	-27.00 [-35.45/8]	-10.95 [-26.11/10]

Table A.12: Lagrangean Relaxation, Part 3

Appendix B

Upper Bounds for the PLSP–MM

This appendix contains the details of the computational results which are summarized in Chapter 6. Five heuristics are tested: Randomized regret based sampling, a cellular automaton, a genetic algorithm, a disjunctive arc based tabu search, and a so–called demand shuffle procedure.

The test–bed as described in Chapter 4 consists of 1,080 small PLSP–MM–instances, 1,033 of them have a feasible solution. It is designed using a full factorial design where the following parameters are systematically varied: The capacity utilization U, the number of machines M, the complexity of the gozinto–structures C, the ratio of setup and holding costs $COSTRATIO$, and the demand pattern $(T_{macro}, T_{micro}, T_{idle})$. For each parameter level combination we have (at most) 10 instances with a feasible solution.

The following tables show three numbers for each parameter level combination: The average deviation from the optimum objective function value, the worst case deviation from the optimum objective function value, and the number of instances for which a feasible solution is found.

B.1 Randomized Regret Based Sampling

U	M	C	$COSTRATIO = 5$ (10,1,5)	(5,2,2)	(1,10,0)
30	1	0.2	16.48 [24.61/10]	13.92 [20.90/10]	1.63 [8.58/10]
30	1	0.8	19.71 [43.71/10]	14.82 [25.02/10]	6.70 [27.64/10]
30	2	0.2	19.95 [36.48/10]	10.18 [22.47/10]	5.63 [12.01/10]
30	2	0.8	20.56 [38.06/9]	13.74 [25.25/10]	7.04 [13.23/10]
50	1	0.2	21.60 [38.42/10]	16.75 [30.93/10]	1.34 [6.74/10]
50	1	0.8	16.95 [34.79/7]	12.57 [26.33/8]	5.96 [16.99/10]
50	2	0.2	24.83 [36.20/10]	11.40 [20.90/10]	8.64 [17.56/10]
50	2	0.8	25.55 [35.88/10]	25.17 [84.92/10]	10.69 [17.78/10]
70	1	0.2	25.89 [42.64/8]	22.88 [39.86/5]	2.80 [8.83/10]
70	1	0.8	11.32 [11.32/1]	3.82 [3.82/1]	7.03 [15.46/10]
70	2	0.2	28.64 [47.42/7]	14.96 [32.05/5]	12.03 [32.03/9]
70	2	0.8	31.08 [49.87/6]	14.02 [17.65/2]	18.71 [26.61/10]

Table B.1: Randomized Regret Based Sampling, Part 1

			$COSTRATIO = 150$		
U	M	C	(10,1,5)	(5,2,2)	(1,10,0)
30	1	0.2	10.78 [16.84/10]	7.01 [12.74/10]	0.08 [0.83/10]
30	1	0.8	14.09 [28.06/10]	13.48 [23.03/10]	2.80 [8.13/10]
30	2	0.2	13.18 [17.86/10]	10.43 [20.67/10]	2.24 [8.89/10]
30	2	0.8	15.15 [26.72/10]	9.79 [16.99/9]	3.09 [8.19/10]
50	1	0.2	12.46 [21.06/10]	8.40 [18.75/10]	0.14 [0.60/10]
50	1	0.8	13.09 [28.17/10]	13.08 [21.65/10]	3.46 [12.18/10]
50	2	0.2	14.02 [20.07/10]	9.02 [14.32/10]	2.97 [10.93/10]
50	2	0.8	15.48 [21.83/10]	22.62 [81.79/10]	3.76 [7.80/10]
70	1	0.2	12.58 [20.04/7]	13.48 [37.02/6]	0.30 [1.66/10]
70	1	0.8	4.89 [4.89/1]	4.80 [5.76/2]	2.29 [5.75/9]
70	2	0.2	16.90 [27.48/7]	17.31 [22.21/5]	1.69 [4.52/9]
70	2	0.8	15.33 [23.74/7]	11.96 [18.49/5]	4.73 [9.65/10]

Table B.2: Randomized Regret Based Sampling, Part 2

			COSTRATIO = 900		
U	M	C	(10,1,5)	(5,2,2)	(1,10,0)
30	1	0.2	5.81 [14.63/10]	3.63 [9.86/10]	0.02 [0.22/10]
30	1	0.8	13.58 [22.76/10]	10.40 [18.22/10]	1.09 [3.07/10]
30	2	0.2	7.94 [13.29/10]	6.05 [16.07/10]	0.55 [1.20/10]
30	2	0.8	15.67 [23.44/9]	10.87 [27.62/9]	1.46 [7.88/10]
50	1	0.2	11.04 [45.73/10]	6.65 [33.74/10]	0.07 [0.33/10]
50	1	0.8	14.39 [20.99/9]	15.65 [34.75/9]	1.05 [2.36/10]
50	2	0.2	7.51 [15.36/10]	6.42 [19.00/9]	1.02 [5.58/10]
50	2	0.8	12.92 [20.21/10]	18.02 [32.57/9]	3.27 [17.38/9]
70	1	0.2	21.30 [49.21/6]	12.51 [22.02/5]	0.16 [1.08/10]
70	1	0.8	— [—/0]	10.54 [12.84/2]	0.20 [0.51/10]
70	2	0.2	16.29 [32.83/7]	19.76 [32.48/4]	0.45 [2.63/9]
70	2	0.8	20.33 [61.32/9]	24.91 [45.51/3]	1.89 [9.90/10]

Table B.3: Randomized Regret Based Sampling, Part 3

B.2 Cellular Automaton

			$COSTRATIO = 5$		
U	M	C	(10,1,5)	(5,2,2)	(1,10,0)
30	1	0.2	16.96 [25.17/10]	16.71 [26.17/10]	1.75 [8.58/10]
30	1	0.8	14.52 [22.12/10]	14.47 [25.92/10]	6.33 [22.05/10]
30	2	0.2	26.93 [47.87/10]	24.33 [44.10/10]	7.12 [24.98/10]
30	2	0.8	25.28 [44.19/9]	25.74 [33.51/10]	8.30 [17.13/10]
50	1	0.2	20.09 [33.58/10]	16.69 [32.20/10]	1.34 [6.74/10]
50	1	0.8	19.55 [31.64/10]	20.56 [43.59/10]	6.74 [18.20/10]
50	2	0.2	36.89 [57.04/10]	26.21 [36.52/9]	8.74 [17.63/10]
50	2	0.8	34.10 [62.36/10]	43.63 [91.67/7]	13.06 [20.45/10]
70	1	0.2	43.90 [78.31/6]	23.77 [35.03/5]	3.39 [8.83/10]
70	1	0.8	12.64 [12.64/1]	11.36 [11.36/1]	7.14 [13.34/10]
70	2	0.2	60.97 [112.78/7]	17.77 [27.17/2]	12.81 [32.03/9]
70	2	0.8	43.98 [60.12/5]	27.64 [27.64/1]	20.62 [31.55/10]

Table B.4: Cellular Automaton, Part 1

			$COSTRATIO = 150$		
U	M	$\mathcal{C}$	(10,1,5)	(5,2,2)	(1,10,0)
30	1	0.2	8.70 [15.89/10]	4.59 [9.86/10]	0.08 [0.83/10]
30	1	0.8	5.22 [13.48/10]	3.94 [10.96/10]	3.09 [8.13/10]
30	2	0.2	13.99 [21.69/10]	9.95 [19.92/10]	2.78 [8.89/10]
30	2	0.8	13.99 [26.78/10]	11.10 [23.86/9]	2.85 [7.64/10]
50	1	0.2	10.76 [21.86/10]	7.28 [17.56/10]	0.14 [0.60/10]
50	1	0.8	13.48 [27.02/10]	6.58 [16.33/10]	3.50 [13.14/10]
50	2	0.2	17.37 [25.52/10]	13.96 [34.02/9]	2.95 [10.95/10]
50	2	0.8	19.83 [29.23/10]	17.96 [25.86/9]	4.47 [8.94/10]
70	1	0.2	27.76 [43.21/5]	20.50 [35.75/4]	0.85 [6.62/10]
70	1	0.8	21.16 [38.00/2]	6.94 [6.94/1]	2.40 [5.75/9]
70	2	0.2	30.12 [44.51/7]	9.52 [13.70/2]	1.82 [4.52/9]
70	2	0.8	27.24 [34.37/3]	28.28 [28.28/1]	5.64 [10.40/10]

Table B.5: Cellular Automaton, Part 2

			COSTRATIO = 900		
U	M	C	(10,1,5)	(5,2,2)	(1,10,0)
30	1	0.2	2.07 [3.64/10]	1.61 [5.04/10]	0.02 [0.22/10]
30	1	0.8	2.08 [6.63/10]	2.18 [5.44/10]	1.11 [3.07/10]
30	2	0.2	4.19 [7.80/10]	3.71 [7.97/10]	0.79 [2.99/10]
30	2	0.8	6.71 [14.74/9]	6.65 [14.90/9]	1.01 [3.74/10]
50	1	0.2	2.70 [7.84/10]	1.44 [3.37/10]	0.03 [0.15/10]
50	1	0.8	3.00 [6.86/10]	7.10 [36.34/9]	1.05 [2.36/10]
50	2	0.2	7.47 [23.72/10]	17.44 [47.58/10]	1.25 [5.75/10]
50	2	0.8	11.52 [22.06/10]	19.53 [38.23/9]	3.61 [18.30/9]
70	1	0.2	24.38 [48.91/5]	12.91 [16.70/3]	0.04 [0.19/10]
70	1	0.8	22.75 [22.75/1]	5.69 [11.28/2]	0.21 [0.51/10]
70	2	0.2	17.97 [55.98/6]	10.66 [19.77/2]	0.29 [1.13/10]
70	2	0.8	22.13 [23.57/3]	— [—/0]	2.45 [13.21/10]

Table B.6: Cellular Automaton, Part 3

B.3 Genetic Algorithm

			$COSTRATIO = 5$		
U	M	C	(10,1,5)	(5,2,2)	(1,10,0)
30	1	0.2	40.72 [73.06/10]	27.19 [43.62/10]	3.61 [13.94/10]
30	1	0.8	53.85 [77.01/10]	27.62 [48.58/10]	9.94 [28.22/10]
30	2	0.2	33.28 [42.36/10]	10.86 [25.60/10]	5.74 [12.01/10]
30	2	0.8	45.64 [83.41/9]	22.82 [35.34/10]	11.03 [22.72/10]
50	1	0.2	49.02 [99.40/10]	34.25 [58.06/10]	4.50 [16.90/10]
50	1	0.8	57.23 [62.06/2]	31.89 [58.44/6]	22.35 [46.62/10]
50	2	0.2	40.15 [81.44/10]	12.01 [21.78/8]	12.17 [22.59/10]
50	2	0.8	48.88 [87.87/9]	33.35 [54.55/9]	17.64 [26.83/10]
70	1	0.2	— [—/0]	39.97 [77.28/6]	8.42 [24.34/10]
70	1	0.8	— [—/0]	21.11 [29.72/3]	29.94 [57.14/10]
70	2	0.2	41.47 [74.65/6]	14.52 [18.23/3]	18.06 [36.42/8]
70	2	0.8	52.44 [67.32/3]	31.57 [48.23/4]	26.69 [45.42/9]

Table B.7: Genetic Algorithm, Part 1

			$COSTRATIO = 150$		
U	M	C	(10,1,5)	(5,2,2)	(1,10,0)
30	1	0.2	27.22 [46.28/10]	14.22 [23.30/10]	1.25 [6.93/10]
30	1	0.8	39.00 [58.50/10]	11.24 [21.95/10]	3.71 [8.17/10]
30	2	0.2	22.80 [38.83/10]	19.63 [36.75/10]	2.66 [7.54/10]
30	2	0.8	36.62 [64.38/10]	18.48 [23.96/9]	5.81 [12.61/10]
50	1	0.2	32.31 [64.29/10]	15.11 [32.50/10]	1.21 [7.81/10]
50	1	0.8	38.81 [40.59/2]	17.50 [31.71/6]	6.41 [15.90/10]
50	2	0.2	25.06 [45.36/10]	16.12 [30.14/8]	4.44 [13.13/10]
50	2	0.8	33.70 [51.29/9]	22.66 [38.13/9]	6.52 [15.14/10]
70	1	0.2	— [—/0]	27.24 [48.40/6]	1.17 [7.44/10]
70	1	0.8	— [—/0]	14.00 [20.06/3]	7.59 [21.80/9]
70	2	0.2	24.17 [40.97/6]	18.07 [26.25/3]	2.70 [9.88/8]
70	2	0.8	37.70 [53.58/3]	18.83 [27.23/4]	5.55 [10.20/9]

Table B.8: Genetic Algorithm, Part 2

			$COSTRATIO = 900$		
U	M	C	(10,1,5)	(5,2,2)	(1,10,0)
30	1	0.2	29.03 [42.00/10]	4.48 [7.20/10]	0.37 [1.98/10]
30	1	0.8	33.40 [44.04/10]	4.71 [9.89/10]	1.70 [4.79/10]
30	2	0.2	26.96 [40.62/10]	36.02 [71.94/10]	1.14 [4.27/10]
30	2	0.8	40.88 [59.30/9]	38.71 [60.22/9]	5.24 [24.63/10]
50	1	0.2	37.19 [70.46/10]	4.34 [7.08/10]	0.27 [1.69/10]
50	1	0.8	43.80 [52.99/2]	10.96 [47.22/6]	1.43 [4.17/10]
50	2	0.2	32.18 [53.89/10]	38.08 [63.37/8]	3.14 [22.41/10]
50	2	0.8	35.82 [47.92/9]	40.85 [48.55/9]	5.71 [20.28/9]
70	1	0.2	— [—/0]	30.91 [43.77/6]	0.05 [0.25/10]
70	1	0.8	— [—/0]	15.72 [30.06/3]	0.31 [1.50/10]
70	2	0.2	35.23 [55.43/6]	45.14 [69.26/3]	0.62 [3.67/8]
70	2	0.8	47.56 [68.85/3]	25.51 [31.49/4]	3.56 [12.51/9]

Table B.9: Genetic Algorithm, Part 3

B.4 Disjunctive Arc Based Tabu Search

			$COSTRATIO = 5$		
U	M	C	(10,1,5)	(5,2,2)	(1,10,0)
30	1	0.2	70.80 [109.62/5]	24.84 [35.10/9]	1.42 [8.58/10]
30	1	0.8	— [—/0]	20.12 [34.37/5]	8.48 [27.64/10]
30	2	0.2	20.66 [40.43/6]	9.13 [22.02/7]	24.03 [51.00/6]
30	2	0.8	29.71 [59.29/7]	11.44 [18.47/8]	14.52 [29.20/7]
50	1	0.2	39.76 [44.52/2]	21.25 [45.46/9]	4.89 [14.60/10]
50	1	0.8	— [—/0]	31.94 [34.68/2]	11.00 [21.23/10]
50	2	0.2	23.28 [57.03/6]	3.82 [11.99/7]	36.58 [65.80/5]
50	2	0.8	29.52 [49.26/7]	9.88 [26.29/8]	28.64 [39.58/7]
70	1	0.2	42.11 [42.11/1]	9.37 [24.98/7]	11.39 [20.27/10]
70	1	0.8	— [—/0]	— [—/0]	16.08 [29.35/10]
70	2	0.2	22.42 [67.99/5]	4.95 [20.62/5]	16.52 [32.65/3]
70	2	0.8	18.61 [38.07/4]	0.61 [3.03/5]	27.78 [45.69/7]

Table B.10: Disjunctive Arc Based Tabu Search, Part 1

U	M	C	$COSTRATIO = 150$ (10,1,5)	(5,2,2)	(1,10,0)
30	1	0.2	43.03 [85.11/9]	17.31 [25.82/10]	0.12 [0.83/10]
30	1	0.8	64.28 [64.28/1]	22.70 [41.02/6]	3.04 [8.13/10]
30	2	0.2	19.06 [28.59/7]	15.91 [27.64/7]	7.78 [17.18/6]
30	2	0.8	22.92 [33.44/8]	13.08 [17.97/7]	6.46 [11.25/7]
50	1	0.2	36.80 [68.22/8]	8.79 [23.14/8]	1.61 [6.88/10]
50	1	0.8	— [—/0]	24.35 [33.75/2]	5.10 [13.14/10]
50	2	0.2	20.95 [58.88/7]	11.70 [30.22/7]	10.48 [25.34/5]
50	2	0.8	21.75 [33.70/8]	6.76 [17.17/8]	10.00 [12.82/7]
70	1	0.2	17.01 [30.02/3]	6.96 [20.16/7]	2.45 [10.91/10]
70	1	0.8	— [—/0]	— [—/0]	5.43 [16.45/9]
70	2	0.2	15.06 [32.31/5]	6.41 [20.34/5]	4.87 [11.15/3]
70	2	0.8	17.07 [32.05/6]	1.28 [6.39/5]	10.66 [19.54/7]

Table B.11: Disjunctive Arc Based Tabu Search, Part 2

			$COSTRATIO = 900$		
U	M	C	(10,1,5)	(5,2,2)	(1,10,0)
30	1	0.2	65.31 [122.60/9]	39.74 [67.17/10]	0.03 [0.22/10]
30	1	0.8	82.86 [82.86/1]	59.68 [78.43/6]	1.09 [3.07/10]
30	2	0.2	47.17 [62.02/7]	33.50 [52.20/7]	2.24 [3.88/6]
30	2	0.8	43.47 [75.66/8]	36.43 [56.76/7]	4.41 [19.42/7]
50	1	0.2	57.96 [79.01/8]	32.55 [60.26/8]	0.44 [1.84/10]
50	1	0.8	— [—/0]	65.23 [80.88/2]	1.31 [2.66/10]
50	2	0.2	46.26 [92.96/7]	21.00 [47.37/7]	2.83 [6.17/5]
50	2	0.8	36.13 [46.79/8]	23.14 [32.54/7]	5.05 [19.54/6]
70	1	0.2	26.86 [41.96/3]	8.03 [18.90/7]	0.40 [3.09/10]
70	1	0.8	— [—/0]	— [—/0]	0.23 [0.52/10]
70	2	0.2	36.90 [58.94/5]	11.42 [43.16/5]	1.17 [2.58/3]
70	2	0.8	24.03 [48.97/6]	7.05 [14.21/5]	2.94 [6.42/7]

Table B.12: Disjunctive Arc Based Tabu Search, Part 3

B.5 Demand Shuffle

			$COSTRATIO = 5$		
U	M	C	(10,1,5)	(5,2,2)	(1,10,0)
30	1	0.2	4.52 [8.45/10]	7.77 [15.64/10]	3.61 [13.94/10]
30	1	0.8	16.07 [30.22/10]	9.26 [21.88/10]	12.55 [33.47/10]
30	2	0.2	7.53 [23.27/10]	6.50 [23.51/10]	4.19 [12.94/10]
30	2	0.8	1.14 [4.41/9]	4.06 [8.23/10]	7.55 [19.44/10]
50	1	0.2	7.25 [13.50/10]	10.08 [25.82/10]	5.47 [22.84/10]
50	1	0.8	13.36 [34.15/10]	9.31 [21.16/10]	16.51 [36.34/10]
50	2	0.2	9.08 [24.55/10]	8.20 [41.76/10]	10.15 [22.59/10]
50	2	0.8	3.13 [8.26/10]	8.33 [16.20/10]	16.30 [37.36/10]
70	1	0.2	9.26 [18.35/10]	6.84 [12.26/8]	7.86 [24.34/10]
70	1	0.8	8.75 [24.74/8]	11.50 [21.36/6]	24.50 [42.93/10]
70	2	0.2	11.82 [25.65/10]	7.58 [19.19/6]	16.10 [36.42/10]
70	2	0.8	5.92 [14.18/9]	6.70 [14.64/7]	26.76 [56.98/10]

Table B.13: Demand Shuffle, Part 1

			$COSTRATIO = 150$		
U	M	C	(10,1,5)	(5,2,2)	(1,10,0)
30	1	0.2	6.64 [20.05/10]	3.19 [9.04/10]	1.25 [6.93/10]
30	1	0.8	12.41 [25.64/10]	5.77 [12.76/10]	4.32 [8.91/10]
30	2	0.2	5.79 [22.24/10]	5.60 [12.88/10]	3.94 [16.21/10]
30	2	0.8	1.87 [5.41/10]	2.53 [4.64/9]	3.59 [7.47/10]
50	1	0.2	5.07 [12.03/10]	3.24 [10.67/10]	1.34 [7.81/10]
50	1	0.8	11.98 [23.32/10]	6.05 [14.04/10]	4.75 [12.77/10]
50	2	0.2	5.17 [11.45/10]	5.91 [19.99/10]	5.74 [25.90/10]
50	2	0.8	2.93 [6.28/10]	6.09 [10.66/10]	4.90 [13.94/10]
70	1	0.2	5.17 [8.68/10]	2.85 [9.88/8]	1.06 [7.44/10]
70	1	0.8	10.21 [21.31/8]	9.77 [26.59/6]	5.54 [21.36/9]
70	2	0.2	7.30 [16.96/10]	7.06 [26.33/6]	4.02 [11.27/10]
70	2	0.8	5.60 [12.07/10]	4.90 [8.19/7]	8.06 [18.02/10]

Table B.14: Demand Shuffle, Part 2

			$COSTRATIO = 900$		
U	M	C	(10,1,5)	(5,2,2)	(1,10,0)
30	1	0.2	2.16 [6.81/10]	0.82 [3.07/10]	0.37 [1.98/10]
30	1	0.8	10.96 [25.20/10]	2.70 [5.44/10]	1.42 [3.20/10]
30	2	0.2	2.77 [6.47/10]	3.33 [12.07/10]	4.97 [24.00/10]
30	2	0.8	4.60 [9.98/9]	2.13 [4.75/9]	6.52 [19.03/10]
50	1	0.2	3.86 [20.08/10]	1.01 [2.71/10]	0.31 [1.69/10]
50	1	0.8	8.74 [21.28/10]	4.10 [21.26/10]	1.16 [2.42/10]
50	2	0.2	3.14 [8.23/10]	9.18 [25.60/10]	7.00 [35.79/10]
50	2	0.8	8.34 [22.11/10]	6.76 [15.92/9]	9.58 [24.12/9]
70	1	0.2	6.34 [19.63/10]	2.07 [6.10/8]	0.02 [0.14/10]
70	1	0.8	7.85 [20.72/8]	1.43 [4.39/6]	0.31 [1.50/10]
70	2	0.2	7.01 [14.31/10]	12.54 [52.69/6]	5.64 [19.73/10]
70	2	0.8	11.08 [29.82/10]	8.98 [16.32/7]	8.79 [32.03/10]

Table B.15: Demand Shuffle, Part 3

Appendix C

PLSP–PM: Demand Shuffle

---

if $(m = 1)$

  for $j = 1$ to $j = J$

    $CD_{jt} := \min\{\ CD_{j(t+1)} + \tilde{d}_{jt},$

      $\max\{0, nr_j - \sum_{\mu \in \mathcal{M}_j} \sum_{\tau=t+1}^{T} q_{j\mu\tau}\}\}.$

    $Q_{jt} := \min\{\ Q_{j(t+1)} + \sum_{\substack{h \in \mathcal{L} \\ item(h)=j \\ deadline(h)=t}} demand(h),$

      $\max\{0, nr_j - \sum_{\mu \in \mathcal{M}_j} \sum_{\tau=t+1}^{T} q_{j\mu\tau}\}\}.$

if $(j_{mt} \neq 0)$

  $q_{j_{mt}mt} := \min\left\{CD_{j_{mt}t}, Q_{j_{mt}t}, \frac{RC_{mt}}{p_{j_{mt}m}}\right\}.$

  $Q_{j_{mt}t} := Q_{j_{mt}t} - q_{j_{mt}mt}.$

  $CD_{j_{mt}t} := CD_{j_{mt}t} - q_{j_{mt}mt}.$

  $RC_{mt} := RC_{mt} - p_{j_{mt}m}q_{j_{mt}mt}.$

  for $i \in \mathcal{P}_{j_{mt}}$

    if $(t - v_i > 0$ and $q_{j_{mt}mt} > 0)$

      $\tilde{d}_{i(t-v_i)} := \tilde{d}_{i(t-v_i)} + a_{ij_{mt}}q_{j_{mt}mt}.$

if $(m = M)$

  DS–PM–construct$(t-1, 1, 1)$.

else

  DS–PM–construct$(t, 0, m+1)$.

Table C.1: Evaluating *DS–PM–construct*$(t, 0, \cdot)$ where $1 \leq t \leq T$

---

choose $j_{mt} \in \mathcal{I}_{mt}$.
if $(j_{mt} \neq 0)$
    $y_{j_{mt}mt} := 1$.
    if $(j_{mt} \neq j_{m(t+1)})$
        $q_{j_{mt}m(t+1)} := \min\left\{CD_{j_{mt}(t+1)}, Q_{j_{mt}(t+1)}, \frac{RC_{m(t+1)}}{p_{j_{mt}m}}\right\}$.
        $Q_{j_{mt}(t+1)} := Q_{j_{mt}(t+1)} - q_{j_{mt}m(t+1)}$.
        $CD_{j_{mt}(t+1)} := CD_{j_{mt}(t+1)} - q_{j_{mt}m(t+1)}$.
        $RC_{m(t+1)} := RC_{m(t+1)} - p_{j_{mt}m}q_{j_{mt}m(t+1)}$.
        for $i \in \mathcal{P}_{j_{mt}}$
            if $(t + 1 - v_i > 0$ and $q_{j_{mt}m(t+1)} > 0)$
                $\tilde{d}_{i(t+1-v_i)} := \tilde{d}_{i(t+1-v_i)} + a_{ij_{mt}}q_{j_{mt}m(t+1)}$.
try to increase the capacity utilization in period $t + 1$.
if $(m = M)$
    DS–PM–construct$(t, 0, 1)$.
else
    DS–PM–construct$(t, 1, m + 1)$.

Table C.2: Evaluating *DS–PM–construct*$(t, 1, \cdot)$ where $1 \leq t < T$

Appendix D

PLSP–MR: Demand Shuffle

$\mathcal{I}_T := \{j \in \{1,\ldots,J\} \mid \tilde{d}_{jT} > 0$
$\quad \wedge \exists h \in \mathcal{L} : (item(h) = j \wedge deadline(h) = T)$
$\quad \wedge \forall m \in \mathcal{M}_j : \sum_{i \in \mathcal{J}_m} y_{imT} = 0\}.$

if $(\mathcal{I}_T \neq \emptyset)$
  choose $j_T \in \mathcal{I}_T$.
  for $m \in \mathcal{M}_{j_T}$
    $j_{mT} := j_T$.
    $y_{j_T mT} := 1$.
  $\mathcal{I}_T := \{i \in \mathcal{I}_T \mid \mathcal{M}_{j_T} \cap \mathcal{M}_i = \emptyset\}$.
  DS–MR–construct$(T, 1)$.
else
  DS–MR–construct$(T, 0)$.

Table D.1: Evaluating *DS–MR–construct*$(T, 1)$

---

for $j \in \{1, \ldots, J\}$

$$CD_{jt} := \min\left\{CD_{j(t+1)} + \tilde{d}_{jt}, \max\{0, nr_j - \sum_{\tau=t+1}^{T}(q_{j\tau}^B + q_{j\tau}^E)\}\right\}.$$

$$Q_{jt} := \min\{ Q_{j(t+1)} + \sum_{\substack{h \in \mathcal{L} \\ item(h)=j \\ deadline(h)=t}} demand(h),$$

$$\max\{0, nr_j - \sum_{\tau=t+1}^{T}(q_{j\tau}^B + q_{j\tau}^E)\}\}.$$

for $j \in \{i \in \{1, \ldots, J\} \mid \forall m \in \mathcal{M}_i : y_{imt} = 1\}$

$$q_{jt}^E := \min\left\{CD_{jt}, Q_{jt}, \min\left\{\frac{RC_{mt}}{p_{jm}} \mid m \in \mathcal{M}_j\right\}\right\}.$$

$$Q_{jt} := Q_{jt} - q_{jt}^E.$$

$$CD_{jt} := CD_{jt} - q_{jt}^E.$$

for $m \in \mathcal{M}_j$

$$RC_{mt} := RC_{mt} - p_{jm}q_{jt}^E.$$

for $i \in \mathcal{P}_j$

if $(t - v_i > 0$ and $q_{jt}^E > 0)$

$$\tilde{d}_{i(t-v_i)} := \tilde{d}_{i(t-v_i)} + a_{ij}q_{jt}^E.$$

$DS\text{-}MR\text{-}construct(t-1, 1)$.

Table D.2: Evaluating $DS\text{-}MR\text{-}construct(t, 0)$ where $1 \leq t \leq T$

$$\mathcal{I}_t := \{j \in \{1,\ldots,J\} \mid CD_{j(t+1)} + \tilde{d}_{jt} > 0 \wedge (Q_{j(t+1)} > 0 \vee \exists h \in \mathcal{L} : (\ item(h) = j \wedge deadline(h) = t)) \wedge \forall m \in \mathcal{M}_j : \sum_{i \in \mathcal{J}_m} y_{imt} = 0\}.$$

if $(\mathcal{I}_t \neq \emptyset)$
    choose $j_t \in \mathcal{I}_t$.
    for $m \in \mathcal{M}_{j_t}$
        $j_{mt} := j_t$.
        $y_{j_t mt} := 1$.
    $\mathcal{I}_t := \{i \in \mathcal{I}_t \mid \mathcal{M}_{j_t} \cap \mathcal{M}_i = \emptyset\}$.
    DS–MR–construct$(t, 1)$.
else
    for $j \in \{i \in \{1,\ldots,J\} \mid \forall m \in \mathcal{M}_i : y_{imt} = 1\}$
        $q^B_{j(t+1)} := \min\{\ CD_{j(t+1)}, Q_{j(t+1)}, \min\left\{\frac{RC_{m(t+1)}}{p_{jm}} \mid m \in \mathcal{M}_j\right\}\}$.
        $Q_{j(t+1)} := Q_{j(t+1)} - q^B_{j(t+1)}$.
        $CD_{j(t+1)} := CD_{j(t+1)} - q^B_{j(t+1)}$.
        for $m \in \mathcal{M}_j$
            $RC_{m(t+1)} := RC_{m(t+1)} - p_{jm} q^B_{j(t+1)}$.
        for $i \in \mathcal{P}_j$
            if $(t + 1 - v_i > 0$ and $q^B_{j(t+1)} > 0)$
                $\tilde{d}_{i(t+1-v_i)} := \tilde{d}_{i(t+1-v_i)} + a_{ij} q^B_{j(t+1)}$.
    try to increase the capacity utilization in period $t + 1$.
    DS–MR–construct$(t, 0)$.

Table D.3: Evaluating *DS–MR–construct*$(t, 1)$ where $1 \leq t < T$

$\mathcal{SI}^B_{(t+1)} := \{j \in \{1,\ldots,J\} \mid \forall m \in \mathcal{M}_j : y_{jmt} = 1\}.$
$\mathcal{SI}^E_{(t+1)} := \left\{j \in \{1,\ldots,J\} \mid \forall m \in \mathcal{M}_j : y_{jm(t+1)} = 1\right\}.$
$\hat{t} := t.$
while ($\exists j \in \mathcal{SI}^B_{(t+1)} \cup \mathcal{SI}^E_{(t+1)} : (CD_{j(t+1)} > Q_{j(t+1)}$
$\wedge \forall m \in \mathcal{M}_j : RC_{m(t+1)} > 0)$
  and $\hat{t} \geq 1$)
  if ($\exists h \in \mathcal{L} :\ item(h) = j$
    $\wedge deadline(h) = \hat{t}$
    $\wedge\ \min\{deadline^I_{UB}(h), deadline^{II}_{UB}(h)\}$
    $\geq t+1$)
    $deadline(h) := t+1.$
    $Q_{j(t+1)} := \min\{\ Q_{j(t+1)} + demand(h),$
    $\max\{0, nr_j - \sum_{\tau=t+1}^{T}(q^B_{j\tau} + q^E_{j\tau})\}\}.$
    $\Delta q_{j(t+1)} := \min\{\ CD_{j(t+1)},$
    $Q_{j(t+1)},$
    $\min\left\{\frac{RC_{m(t+1)}}{p_{jm}} \mid m \in \mathcal{M}_j\right\}\}.$
    if ($j \in \mathcal{SI}^B_{(t+1)}$)
      $q^B_{j(t+1)} := q^B_{j(t+1)} + \Delta q_{j(t+1)}.$
    else
      $q^E_{j(t+1)} := q^E_{j(t+1)} + \Delta q_{j(t+1)}.$
    $Q_{j(t+1)} := Q_{j(t+1)} - \Delta q_{j(t+1)}.$
    $CD_{j(t+1)} := CD_{j(t+1)} - \Delta q_{j(t+1)}.$
    for $m \in \mathcal{M}_j$
      $RC_{m(t+1)} := RC_{m(t+1)} - p_{jm}\Delta q_{j(t+1)}.$
    for $i \in \mathcal{P}_j$
      if ($t+1-v_i > 0$ and $\Delta q_{j(t+1)} > 0$)
        $\tilde{d}_{i(t+1-v_i)} := \tilde{d}_{i(t+1-v_i)} + a_{ij}\Delta q_{j(t+1)}.$
  else
    $\hat{t} := \hat{t} - 1.$

Table D.4: A Procedure to Increase the Capacity Utilization in Period $t+1$

$\mathcal{I}_0 := \{j \in \{1, \ldots, J\} \mid CD_{j1} > 0 \wedge Q_{j1} > 0 \wedge \forall m \in \mathcal{M}_j : y_{jm0} = 1\}$.
for $j \in \mathcal{I}_0$

$\quad q_{j1}^B := \min\left\{CD_{j1}, Q_{j1}, \min\left\{\frac{RC_{m1}}{p_{jm}} \mid m \in \mathcal{M}_j\right\}\right\}$.

$\quad Q_{j1} := Q_{j1} - q_{j1}^B$.

$\quad CD_{j1} := CD_{j1} - q_{j1}^B$.

Table D.5: Evaluating *DS–MR–construct*$(0, 1)$

Appendix E

PLSP–PRR: Demand Shuffle

---

for $j \in \mathcal{J}_m$

  $CD_{jt} := \min\left\{CD_{j(t+1)} + \tilde{d}_{jt}, \max\{0, nr_j - \sum_{\tau=t+1}^{T} q_{j\tau}\}\right\}.$

  $Q_{jt} := \min\{ Q_{j(t+1)} + \sum_{\substack{h \in \mathcal{L} \\ item(h)=j \\ deadline(h)=t}} demand(h),$

  $\max\{0, nr_j - \sum_{\tau=t+1}^{T} q_{j\tau}\}\}.$

if $(j_{mt} \neq 0)$

  $q_{j_{mt}t} := \min\{ CD_{j_{mt}t},$

  $Q_{j_{mt}t},$

  $\frac{RC_{mt}}{p_{j_{mt}}},$

  $\min\left\{\frac{\tilde{RC}_{\mu\lceil\frac{t}{L}\rceil}}{\tilde{p}_{j_{mt}\mu}} \mid \mu \in \tilde{\mathcal{M}}_{j_{mt}}\right\}\}.$

  $Q_{j_{mt}t} := Q_{j_{mt}t} - q_{j_{mt}t}.$

  $CD_{j_{mt}t} := CD_{j_{mt}t} - q_{j_{mt}t}.$

  $RC_{mt} := RC_{mt} - p_{j_{mt}} q_{j_{mt}t}.$

  for $\mu \in \tilde{\mathcal{M}}_{j_{mt}}$

    $\tilde{RC}_{\mu\lceil\frac{t}{L}\rceil} := \tilde{RC}_{\mu\lceil\frac{t}{L}\rceil} - \tilde{p}_{j_{mt}\mu} q_{j_{mt}t}.$

  for $i \in \mathcal{P}_{j_{mt}}$

    if $(t - v_i > 0$ and $q_{j_{mt}t} > 0)$

      $\tilde{d}_{i(t-v_i)} := \tilde{d}_{i(t-v_i)} + a_{ij_{mt}} q_{j_{mt}t}.$

if $(m = M)$

  DS–PRR–construct$(t-1, 1, 1)$.

else

  DS–PRR–construct$(t, 0, m+1)$.

Table E.1: Evaluating *DS–PRR–construct*$(t, 0, \cdot)$ where $1 \leq t \leq T$

---

choose $j_{mt} \in \mathcal{I}_{mt}$.
if $(j_{mt} \neq 0)$
    $y_{j_{mt}t} := 1$.
    if $(j_{mt} \neq j_{m(t+1)})$
        $$q_{j_{mt}(t+1)} := \min\{\ CD_{j_{mt}(t+1)},\ Q_{j_{mt}(t+1)},\ \frac{RC_{m(t+1)}}{p_{j_{mt}}},\ \min\left\{ \frac{\tilde{RC}_{\mu\lceil\frac{t+1}{L}\rceil}}{\tilde{p}_{j_{mt}\mu}} \mid \mu \in \tilde{\mathcal{M}}_{j_{mt}} \right\}\}.$$
        $Q_{j_{mt}(t+1)} := Q_{j_{mt}(t+1)} - q_{j_{mt}(t+1)}$.
        $CD_{j_{mt}(t+1)} := CD_{j_{mt}(t+1)} - q_{j_{mt}(t+1)}$.
        $RC_{m(t+1)} := RC_{m(t+1)} - p_{j_{mt}} q_{j_{mt}(t+1)}$.
        for $\mu \in \tilde{\mathcal{M}}_{j_{mt}}$
            $\tilde{RC}_{\mu\lceil\frac{t+1}{L}\rceil} := \tilde{RC}_{\mu\lceil\frac{t+1}{L}\rceil} - \tilde{p}_{j_{mt}\mu} q_{j_{mt}(t+1)}$.
        for $i \in \mathcal{P}_{j_{mt}}$
            if $(t+1-v_i > 0$ and $q_{j_{mt}(t+1)} > 0)$
                $\tilde{d}_{i(t+1-v_i)} := \tilde{d}_{i(t+1-v_i)} + a_{ij_{mt}} q_{j_{mt}(t+1)}$.
try to increase the capacity utilization in period $t+1$.
if $(m = M)$
    DS–PRR–construct$(t, 0, 1)$.
else
    DS–PRR–construct$(t, 1, m+1)$.

Table E.2: Evaluating *DS–PRR–construct*$(t, 1, \cdot)$ where $1 \leq t < T$

$\hat{t} := t.$

while ($(CD_{j_{mt}(t+1)} > Q_{j_{mt}(t+1)}$ or $CD_{j_{m(t+1)}(t+1)} > Q_{j_{m(t+1)}(t+1)})$
  and $RC_{m(t+1)} > 0$
  and ($\forall \mu \in \tilde{\mathcal{M}}_{j_{mt}} : \tilde{RC}_{\mu(t+1)} > 0$
    or $\forall \mu \in \tilde{\mathcal{M}}_{j_{m(t+1)}} : \tilde{RC}_{\mu(t+1)} > 0)$
  and $\hat{t} \geq 1)$

 if $(\exists h \in \mathcal{L} : (item(h) = j_{mt} \vee item(h) = j_{m(t+1)})$
   $\wedge deadline(h) = \hat{t}$
   $\wedge \min\{deadline^{I}_{UB}(h), deadline^{II}_{UB}(h)\}$
    $\geq t+1$
   $\wedge CD_{item(h)(t+1)} > Q_{item(h)(t+1)})$

  $deadline(h) := t+1.$

  $$Q_{item(h)(t+1)} := \min\{ Q_{item(h)(t+1)} + demand(h), \max\{0, nr_{item(h)} - \sum_{\tau=t+1}^{T} q_{item(h)\tau}\}\}.$$

  $$\Delta q_{item(h)(t+1)} := \min\left\{ CD_{item(h)(t+1)}, Q_{item(h)(t+1)}, \frac{RC_{m(t+1)}}{p_{item(h)}}, \min\left\{ \frac{\tilde{RC}_{\mu\lceil\frac{t+1}{L}\rceil}}{\tilde{p}_{item(h)\mu}} \mid \mu \in \tilde{\mathcal{M}}_{item(h)} \right\}\right\}.$$

  $q_{item(h)(t+1)} := q_{item(h)(t+1)} + \Delta q_{item(h)(t+1)}.$
  $Q_{item(h)(t+1)} := Q_{item(h)(t+1)} - \Delta q_{item(h)(t+1)}.$
  $CD_{item(h)(t+1)} := CD_{item(h)(t+1)} - \Delta q_{item(h)(t+1)}.$
  $RC_{m(t+1)} := RC_{m(t+1)} - p_{item(h)}\Delta q_{item(h)(t+1)}.$
  for $\mu \in \tilde{\mathcal{M}}_{item(h)}$
   $\tilde{RC}_{\mu\lceil\frac{t+1}{L}\rceil} := \tilde{RC}_{\mu\lceil\frac{t+1}{L}\rceil} - \tilde{p}_{item(h)\mu}\Delta q_{item(h)(t+1)}.$
  for $i \in \mathcal{P}_{item(h)}$
   if $(t+1-v_i > 0$ and $\Delta q_{item(h)(t+1)} > 0)$
    $\tilde{d}_{i(t+1-v_i)} := \tilde{d}_{i(t+1-v_i)} + a_{i(item(h))}\Delta q_{item(h)(t+1)}.$

 else
  $\hat{t} := \hat{t} - 1.$

Table E.3: A Procedure to Increase the Capacity Utilization in Period $t+1$

Bibliography

[AaPl94] AAMODT, A., PLAZA, E., (1994), Case–Based Reasoning: Foundational Issues, Methodological Variations, and System Approaches, AI Communications, Vol. 7, pp. 39–59

[And29] ANDLER, K., (1929), Rationalisierung der Fabrikation und optimale Losgröße, München, Oldenbourg

[Afe85] AFENTAKIS, P., (1985), Simultaneous Lot Sizing and Sequencing for Multistage Production Systems, IIE Transactions, Vol. 17, pp. 327–331

[Afe87] AFENTAKIS, P., (1987), A Parallel Heuristic Algorithm for Lot–Sizing in Multi–Stage Production Systems, IIE Transactions, Vol. 19, pp. 34–42

[AfGa86] AFENTAKIS, P., GAVISH, B., (1986), Optimal Lot–Sizing Algorithms for Complex Product Structures, Operations Research, Vol. 34, pp. 237–249

[AfGaKa84] AFENTAKIS, P., GAVISH, B., KARMARKAR, U., (1984), Computationally Efficient Optimal Solutions to the Lot–Sizing Problem in Multistage Assembly Systems, Management Science, Vol. 30, pp. 222–239

[AgPa93] AGGARWAL, A., PARK, J.K., (1993), Improved Algorithms for Economic Lot–Size Problems, Operations Research, Vol. 41, pp. 549–571

[AkEr88] AKSOY, Y., ERENGUC, S.S., (1988), Multi–Item Inventory Models with Coordinated Replenishments: A Survey, International Journal of Production Research, Vol. 22, pp. 923–935

[AlBa96] ALTHOFF, K.D., BARTSCH–SPÖRL, B., (1996), Decision Support for Case–Based Applications, Wirtschaftsinformatik, Vol. 38, pp. 8–16

[APICS95a] AMERICAN PRODUCTION AND INVENTORY CONTROL SOCIETY INC., (1995), APICS Dictionary, Falls Church, 8th edition

[APICS95b] AMERICAN PRODUCTION AND INVENTORY CONTROL SOCIETY INC., (1995), MRP II Software/Vendor Directory, APICS — The Performance Advantage, No. 9, pp. 38–48

[APICS95c] AMERICAN PRODUCTION AND INVENTORY CONTROL SOCIETY INC., (1995), MRP II Software/Vendor Directory Addendum, APICS — The Performance Advantage, No. 11, pp. 54–56

[ArJoRo89] ARKIN, E., JONEJA, D., ROUNDY, R., (1989), Computational Complexity of Uncapacitated Multi–Echelon Production Planning Problems, Operations Research Letters, Vol. 8, pp. 61–66

[Atk94] ATKINS, D.R., (1994), A Simple Lower Bound to the Dynamic Assembly Problem, European Journal of Operational Research, Vol. 75, pp. 462–466

[AtQuSu92] ATKINS, D.R., QUEYRANNE, M., SUN, D., (1992), Lot Sizing Policies for Finite Production Rate Assembly Systems, Operations Research, Vol. 40, pp. 126–141

[AtSu95] ATKINS, D. R., SUN, D., (1995), 98%–Effective Lot Sizing for Series Inventory Systems with Backlogging, Operations Research, Vol. 43, pp. 335–345

[AxNu86] AXSÄTER, S., NUTTLE, H.L.W., (1986), Aggregating Items in Multi–Level Lot Sizing, in: Axsäter, S., Schneeweiß, C, Silver, E., (eds.), Multi–Stage Production Planning and Inventory Control, Lecture Notes in Economics and Mathematical Systems, Vol. 266, Springer, Berlin, pp. 109–118

[AxNu87] AXSÄTER, S., NUTTLE, H.L.W., (1987), Combining Items for Lot Sizing in Multi–Level Assembly Systems, International Journal of Production Research, Vol. 25, pp. 795–807

[AxRo94] AXSÄTER, S., ROSLING, K., (1994), Multi–Level Production–Inventory Control: Material Requirements Planning or Reorder Point Policies?, European Journal of Operational Research, Vol. 75, pp. 405–412

[BaRi84] BAHL, H.C., RITZMAN, L.P., (1984), An Integrated Model for Master Scheduling, Lot Sizing and Capacity Requirements Planning, Journal of the Operational Research Society, Vol. 35, pp. 389–399

[BaRiGu87] BAHL, H.C., RITZMAN, L.P., GUPTA, J.N.D., (1987), Determining Lot Sizes and Resource Requirements: A Review, Operations Research, Vol. 35, pp. 329–345

[Bak74] BAKER, K.R., (1974), Introduction to Sequencing and Scheduling, New York, John Wiley & Sons

[Bak77] BAKER, K.R., (1977), An Experimental Study of the Effectiveness of Rolling Schedules in Production Planning, Decision Sciences, Vol. 8, pp. 19–27

[Bak81] BAKER, K.R., (1981), An Analysis of Terminal Conditions in Rolling Schedules, European Journal of Operational Research, Vol. 7, pp. 355–361

[Bak90] BAKER, K.R., (1990), Lot–Sizing Procedures and a Standard Data Set: A Reconsiliation of the Literature, Journal of Manufacturing and Operations Management, Vol. 2, pp. 199–221

[BaPe79] BAKER, K.R., PETERSON, D.W., (1979), An Analytic Framework for Evaluating Rolling Schedules, Management Science, Vol. 25, pp. 341–351

[Bal69] BALAS, E., (1969), Machine Sequencing via Disjunctive Graphs: An Implicit Enumeration Algorithm, Operations Research, Vol. 17, pp. 941–957

[BaCo92] BAMBERG, G., COENENBERG, A.G., (1992), Betriebswirtschaftliche Entscheidungslehre, München, Vahlen, 7th edition

[BaGoKeReSt95] BARR, R.S., GOLDEN, B.L., KELLY, J.P., RESENDE, M.G.C., STEWART, W.R., (1995), Designing and Reporting on Computational Experiments with Heuristic Methods, Journal of Heuristics, Vol. 1, pp. 9–32

[Bau52] BAUMOL, W.J., (1952), The Transactions Demand for Cash: An Inventory Theoretic Approach, Journal of Economics, Vol. 66, pp. 545–556

[BeSmYa87] BEAN, J., SMITH, R., YANO, C., (1987), Forecast Horizons for the Discounted Dynamic Lot Size Model Allowing Speculative Motive, Naval Research Logistics, Vol. 34, pp. 761–774

[BeBrPu85] VAN BEEK, P., BREMER, A., VAN PUTTEN, C., (1985), Design and Optimization of Multi–Echelon Assembly Networks: Savings and Potentialities, European Journal of Operational Research, Vol. 19, pp. 57–67

[BeCrPr83] BENSOUSSAN, A., CROUHY, J., PROTH, J.M., (1983), Mathematical Theory of Production Planning, Amsterdam, North–Holland

[BeSr85] BENTON, W.C., SRIVASTAVA, R., (1985), Product Structure Complexity and Multilevel Lot Sizing Using Alternative Costing Policies, Decision Sciences, Vol. 16, pp. 357–369

[Ber72] BERRY, W.L., (1972), Lot Sizing Procedures for Material Requirements Planning Systems: A Framework for Analysis, Production and Inventory Management Journal, Vol. 13, No. 2, pp. 19–35

[BeDoPa81] BEY, R.P., DOERSCH, R.H., PATTERSON, J.H., (1981), The Net Present Value Criterion: Its Impact on Project Scheduling, Project Management Quarterly, Vol. 12, pp. 35–45

[Big75] BIGGS, J.R., (1975), An Analysis of Heuristic Lot Sizing and Sequencing Rules on the Performance of a Hierarchical Multi–Product, Multi–Stage Production Inventory System Utilizing Material Requirements Planning, Ph.D. dissertation, Ohio State University

[Big79] BIGGS, J.R., (1979), Heuristic Lot–Sizing and Sequencing Rules in a Multistage Production–Inventory System, Decision Sciences, Vol. 10, pp. 96–115

[Big85] BIGGS, J.R., (1985), Priority Rules for Shop Floor Control in a Material Requirements Planning System under Various Levels of Capacity, International Journal of Production Research, Vol. 23, pp. 33–46

[BiGoHa77] BIGGS, J.R., GOODMAN, S.H., HARDY, S.T., (1977), Lot Sizing Rules in a Hierarchical Multi–Stage Inventory System, Production and Inventory Management Journal, Vol. 18, No. 1, pp. 104–115

[BiHaPi80] BIGGS, J.R., HAHN, C.K., PINTO, P.A., (1980), Performance of Lot–Sizing Rules in an MRP System with Different Operating Conditions, Academy of Management Review, Vol. 5, pp. 89–96

[BiMo79] BIGHAM, P.E., MOGG, J.M., (1979), Converging–Branch, Multi–Stage Production Schedules with Finite Production Rates and Startup Delays, Journal of the Operational Research Society, Vol. 30, pp. 737–745

[Bil83] BILLINGTON, P.J., (1983), Multi–Level Lot–Sizing with a Bottleneck Work Center, Ph.D. dissertation, Cornell University

[BiMClTh83] BILLINGTON, P.J., MCCLAIN, J.O., THOMAS, L.J., (1983), Mathematical Programming Approaches to Capacity-Constrained MRP Systems: Review, Formulation and Problem Reduction, Management Science, Vol. 29, pp. 1126–1141

[BiMClTh86] BILLINGTON, P.J., MCCLAIN, J.O., THOMAS, L.J., (1986), Heuristics for Multilevel Lot-Sizing with a Bottleneck, Management Science, Vol. 32, pp. 989–1006

[BiHaHa83] BITRAN, G.R., HAAS, E.A., HAX, A.C., (1983), Hierarchical Production Planning: A Two-Stage System, Operations Research, Vol. 30, pp. 232–251

[BiMa86] BITRAN, G.R., MATSUO, H., (1986), Approximation Formulations for the Single-Product Capacitated Lot Size Problem, Operations Research, Vol. 34, pp. 63–74

[BiYa82] BITRAN, G.R., YANASSE, H.H., (1982), Computational Complexity of the Capacitated Lot Size Problem, Management Science, Vol. 28, pp. 1174–1186

[Bit81] BITZ, M., (1981), Entscheidungstheorie, München, Vahlen

[BlKu74] BLACKBURN, J.D., KUNREUTHER, H., (1974), Planning Horizons for the Dynamic Lot Size Model with Backlogging, Management Science, Vol. 21, pp. 251–255

[BlMi80] BLACKBURN, J.D., MILLEN, R.A., (1980), Heuristic Lot-Sizing Performance in a Rolling-Schedule Environment, Decision Sciences, Vol. 11, pp. 691–701

[BlMi82a] BLACKBURN, J.D., MILLEN, R.A., (1982), Improved Heuristics for Multi-Stage Requirements Planning Systems, Management Science, Vol. 28, pp. 44–56

[BlMi82b] BLACKBURN, J.D., MILLEN, R.A., (1982), The Impact of a Rolling Schedule in Multi-Level MRP Systems, Journal of Operations Management, Vol. 2, pp. 125–135

[BlMi84] BLACKBURN, J.D., MILLEN, R.A., (1984), Simultaneous Lot-Sizing and Capacity Planning in Multi-Stage Assembly Processes, European Journal of Operational Research, Vol. 16, pp. 84–93

[BlMi85] BLACKBURN, J.D., MILLEN, R.A., (1985), An Evaluation of Heuristic Performance in Multi-Stage Lot-Sizing Systems, International Journal of Production Research, Vol. 23, pp. 857–866

[BlKrMi86] BLACKBURN, J.D., KROPP, D.H., MILLEN, R.A., (1986), A Comparison of Strategies to Dampen Nervousness in MRP Systems, Management Science, Vol. 32, pp. 413–429

[BlKrMi87] BLACKBURN, J.D., KROPP, D.H., MILLEN, R.A., (1987), Alternative Approaches to Schedule Instability: A Comparative Analysis, International Journal of Production Research, Vol. 25, pp. 1739–1749

[Bla83] BLAHUT, R.E., (1983), Theory and Practice of Error Control Codes, Reading, Addison–Wesley

[Bla87] BŁAŻEWICZ, J., (1987), Selected Topics in Scheduling Theory, Annals of Discrete Mathematics, Vol. 31, pp. 1–59

[BlCeSlWe86] BŁAŻEWICZ, J., CELLARY, W., SLOWINSKI, R., WEGLARZ, J., (1986), Scheduling Under Resource Constraints: Deterministic Models, Annals of Operations Research, Vol. 7, pp. 11–132

[BlDrWe91] BŁAŻEWICZ, J., DROR, M., WEGLARZ, J., (1991), Mathematical Programming Formulations for Machine Scheduling: A Survey, European Journal of Operational Research, Vol. 51, pp. 283–300

[BoKo90] BOOKBINDER, J.H., KOCH, L.A., (1990), Production Planning for Mixed Assembly / Arborescent Systems, Journal of Operations Management, Vol. 9, pp. 7–23

[BrMy88] BREALEY, R.A., MYERS, S.E., (1988), Principles of Corporate Finance, New York, McGraw–Hill, 3rd edition

[Bru95] BRUCKER, P., (1995), Scheduling Algorithms, Berlin, Springer

[Brü95] BRÜGGEMANN, W., (1995), Ausgewählte Probleme der Produktionsplanung — Modellierung, Komplexität und neuere Lösungsmöglichkeiten, Physica–Schriften zur Betriebswirtschaft, Vol. 52, Heidelberg, Physica

[BrJa94] BRÜGGEMANN, W., JAHNKE, H., (1994), DLSP for 2–Stage Multi–Item Batch Production, International Journal of Production Research, Vol. 32, pp. 755–768

[Cam92a] CAMPBELL, G.M., (1992), Using Short–Term Dedication for Scheduling Multiple Products on Parallel Machines, Production and Operations Management, Vol. 1, pp. 295–307

[Cam92b] CAMPBELL, G.M., (1992), Master Production Scheduling under Rolling Planning Horizon with Fixed Order Intervals, Decision Sciences, Vol. 23, pp. 312–331

[CaBeKr82] CARLSON, R.C., BECKMAN, S.L., KROPP, D.H., (1982), The Effectiveness of Extending the Horizon in Rolling Production Scheduling, Decision Sciences, Vol. 13, pp. 129–146

[CaJuKr79] CARLSON, R.C., JUCKER, J.V., KROPP, D.H., (1979), Less Nervous MRP Systems: A Dynamic Economic Lot–Sizing–Approach, Management Science, Vol. 25, pp. 754–761

[Car90] CARRENO, J.J., (1990), Economic Lot Scheduling for Multiple Products on Parallel Identical Processors, Management Science, Vol. 36, pp. 348–358

[CaMaWa90] CATTRYSSE, D., MAES, J., VAN WASSENHOVE, L.N., (1990), Set Partitioning and Column Generation Heuristics for Capacitated Dynamic Lotsizing, European Journal of Operational Research, Vol. 46, pp. 38–47

[CaSaKuWa93] CATTRYSSE, D., SALOMON, M., KUIK, R., VAN WASSENHOVE, L.N., (1993), A Dual Ascent and Column Generation Heuristic for the Discrete Lotsizing and Scheduling Problem, Management Science, Vol. 39, pp. 477–486

[Cha84] CHAKRAVARTY, A.K., (1984), Multi–Stage Production/Inventory Deterministic Lot Size Computations, International Journal of Production Research, Vol. 22, pp. 405–420

[ChSh85] CHAKRAVARTY, A.K., SHTUB, A., (1985), An Experimental Study of the Efficiency of Integer Multiple Lot Sizes in Multi–Echelon Production Inventory Systems, International Journal of Production Research, Vol. 23, pp. 469–478

[Cha82] CHAND, S., (1982), A Note on Dynamic Lot Sizing in a Rolling–Horizon Environment, Decision Sciences, Vol. 13, pp. 113–119

[Cha89] CHAND, S., (1989), Lot Sizes and Setup Frequency with Learning in Setups and Process Quality, European Journal of Operational Research, Vol. 42, pp. 190–202

[ChMo86] CHAND, S., MORTON, T.E., (1986), Minimal Forecast Horizon Procedures for Dynamic Lot Size Models, Naval Research Logistics, Vol. 33, pp. 111–122

[ChMoPr90] CHAND, S., MORTON, T.E., PROTH, J.M., (1990), Existence of Forecast Horizons in Undiscounted Discrete-Time Lot Size Models, Operations Research, Vol. 38, pp. 884–892

[ChSe90] CHAND, S., SETHI, S.P., (1990), A Dynamic Lot Sizing Model with Learning in Setups, Operations Research, Vol. 38, pp. 644–655

[ChFi94] CHANDRA, P., FISHER, M.L., (1994), Coordination of Production and Distribution Planning, European Journal of Operational Research, Vol. 72, pp. 503–517

[ChZh94] CHEN, F., ZHENG, Y.S., (1994), Lower Bounds for Multi-Echelon Stochastic Inventory Systems, Management Science, Vol. 40, pp. 1426–1443

[ChHeLe95] CHEN, H.D., HEARN, D.W., LEE, C.Y., (1995), Minimizing the Error Bound for the Dynamic Lot Size Model, Operations Research Letters, Vol. 17, pp. 57-68

[ChSi90] CHENG, T.C.E., SIN, C.C.S., (1990), A State-of-the-Art Review of Parallel-Machine Scheduling Research, European Journal of Operational Research, Vol. 47, pp. 271–292

[ChLi89] CHIU, H.N., LIN, T.M., (1989), An Optimal Model and a Heuristic Technique for Multi-Stage Lot-Sizing Problems: Algorithms and Performance Tests, Engineering Costs and Production Economics, Vol. 16, pp. 151–160

[ChMaCl84] CHOI, H.G., MALSTROM, E.M., CLASSEN, R. J., (1984), Computer Simulation of Lot-Sizing Algorithms in Three-Stage Multi-Echelon Inventory Systems, Journal of Operations Management, Vol. 4, No. 4, pp. 259–278

[ChMaTs88] CHOI, H.G., MALSTROM, E.M., TSAI, R.D., (1988), Evaluating Lot-Sizing Methods in Multilevel Inventory Systems by Simulation, Production and Inventory Management Journal, Vol. 29, pp. 4–10

[ChCh90] CHOO, E.U., CHAN, G.H., (1990), Two Way Eyeballing Heuristics in Dynamic Lot Sizing with Backlogging, Computers & Operations Research, Vol. 17, pp. 359–363

[ChFlLi94] CHUNG, C.S., FLYNN, J., LIN, C.H.M., (1994), An Effective Algorithm for the Capacitated Single Item Lot Size Problem, European Journal of Operational Research, Vol. 75, pp. 427–440

[ChMe94] CHUNG, C.S., MERCAN, H.M., (1994), Heuristic Procedure for a Multiproduct Dynamic Lot-Sizing Problem with Coordinated Replenishments, Naval Research Logistics, Vol. 41, pp. 273–286

[Cla72] CLARK, A.J., (1972), An Informal Survey of Multi-Echelon Inventory Theory, Naval Research Logistics, Vol. 19, pp. 621–650

[ClSc60] CLARK, A.J., SCARF, H., (1960), Optimal Policies for a Multi-Echelon Inventory Problem, Management Science, Vol. 6, pp. 475–490

[ClAr95] CLARK, A.R., ARMENTANO, V.A., (1995), A Heuristic for a Resource-Capacitated Multi-Stage Lot Sizing Problem with Lead Times, Journal of the Operational Research Society, Vol. 46, pp. 1208–1222

[CoVa82] COLE, F., VANDERVEKEN, H., (1982), Deterministic Lot-Sizing Procedures for Multi-Stage Assembly Systems, Engineering Cost and Production Economics, Vol. 7, pp. 37–44

[CoMKn90] COLEMAN, B.J., MCKNEW, M.A., (1990), A Technique for Order Placement and Sizing, Journal of Purchasing and Materials Management, Vol. 26, pp. 32–40

[CoMKn91] COLEMAN, B.J., MCKNEW, M.A., (1991), An Improved Heuristic for Multilevel Lot Sizing in Material Requirements Planning, Decision Sciences, Vol. 22, pp. 136–156

[Col80a] COLLIER, D.A., (1980), The Interaction of Single-Stage Lot Size Models in a Material Requirements Planning System, Production and Inventory Management, Vol. 4, pp. 11–20

[Col80b] COLLIER, D.A., (1980), A Comparison of MRP Lot-Sizing Methods Considering Capacity Change Costs, Journal of Operations Management, Vol. 1, pp. 23–29

[Col81] COLLIER, D.A., (1981), The Measurement and Operating Benefits of Component Part Commonality, Decision Sciences, Vol. 12, pp. 85–96

[Col82a] COLLIER, D.A., (1982), Aggregate Safety Stock Levels and Component Part Commonality, Management Science, Vol. 28, pp. 1296–1303

[Col82b] COLLIER, D.A., (1982), Research Issues for Multi-Level Lot Sizing MRP Systems, Journal of Operations Management, Vol. 2, pp. 113–123

[CoMaMi67] CONWAY, R.W., MAXWELL, W.L., MILLER, L.W., (1967), Theory of Scheduling, Reading, Addison–Wesley

[CoKu89] COOLEN, A.C.C., KUJIK, F.W., (1989), A Learning Mechanism for Invariant Pattern Recognition in Neural Networks, Neural Networks, Vol. 2, pp. 495–506

[CrWa73] CROWSTON, W.B., WAGNER, M.H., (1973), Dynamic Lot Size Models for Multi–Stage Assembly Systems, Management Science, Vol. 20, pp. 14–21

[CrWaHe72] CROWSTON, W.B., WAGNER, M.H., HENSHAW, A., (1972), A Comparison of Exact and Heuristic Routines for Lot–Size Determination in Multi–Stage Assembly Systems, AIIE Transactions, Vol. 4, pp. 313–317

[CrWaWi73] CROWSTON, W.B., WAGNER, M., WILLIAMS, J.F., (1973), Economic Lot Size Determination in Multi–Stage Assembly Systems, Management Science, Vol. 19, pp. 517–527

[DaMe85] DAGLI, C.H., MERAL, F.S., (1985), Multi–Level Lot–Sizing Heuristics, in: Bullinger, H.J., Warnecke, H.J., (eds.), Towards the Factory of the Future, Berlin, Springer

[DaKo95] DANIELS, R.L., KOUVELIS, P., (1995), Robust Scheduling to Hedge Against Processing Time Uncertainty in Single–Stage Production, Management Science, Vol. 41, pp. 363–376

[DaLa94a] DAUZÈRE–PÉRES, S., LASSERRE, J.B., (1994), Integration of Lotsizing and Scheduling Decisions in a Job-Shop, European Journal of Operational Research, Vol. 75, pp. 413–426

[DaLa94b] DAUZÈRE–PÉRES, S., LASSERRE, J.B., (1994), An Integrated Approach in Production Planning and Scheduling, Lecture Notes in Economics and Mathematical Systems, Berlin, Springer, Vol. 411

[Dav66] DAVIS, E.W., (1966), Resource Allocation in Project Network Models — A Survey, Journal of Industrial Engineering, Vol. 17, pp. 177–188

[Dav73] DAVIS, E.W., (1973), Project Scheduling under Resource Constraints — Historical Review and Categorization of Procedures, AIIE Transactions, Vol. 5, pp. 297–313

[DaPa75] DAVIS, E.W., PATTERSON, J.H., (1975), A Comparison of Heuristic and Optimum Solutions in Resource–Constrained Project Scheduling, Management Science, Vol. 21, pp. 944–955

[Dav91] DAVIS, L., (ed.), (1991), Handbook of Genetic Algorithms, New York, van Nostrand Reinhold

[DBoGeWa84] DEBOTH, M.A., GELDERS, L.F., VAN WASSENHOVE, L.N., (1984), Lot Sizing Under Dynamic Demand Conditions: A Review, Engineering Costs and Production Economics, Vol. 8, pp. 165–187

[DBoWa83] DEBOTH, M.A., VAN WASSENHOVE, L.N., (1983), Cost Increases due to Demand Uncertainty in MRP Lot Sizing, Decision Sciences, Vol. 14, pp. 345–361

[DBoWaGe82] DEBOTH, M.A., VAN WASSENHOVE, L.N., GELDERS, L.F., (1982), Lot–Sizing and Safety Stock Decision in an MRP with Demand Uncertainty, Engineering Costs and Production Economics, Vol. 6, pp. 67–75

[Dem92] DEMEULEMEESTER, E., (1992), Optimal Algorithms for Various Classes of Multiple Resource–Constrained Project Scheduling Problems, Ph.D. dissertation, Katholieke Universiteit Leuven

[DeLe91] DENARDO, E.V., LEE, C.Y., (1991), Error Bound for the Dynamic Lot Size Model with Backlogging, Annals of Operations Research, Vol. 2, pp. 213–228

[Der95] DERSTROFF, M., (1995), Mehrstufige Losgrößenplanung mit Kapazitätsbeschränkungen, Heidelberg, Physica

[DiBaKaZi92a] DIABY, M., BAHL, H.C., KARWAN, M.H., ZIONTS, S., (1992), A Lagrangean Relaxation Approach for Very–Large–Scale Capacitated Lot–Sizing, Management Science, Vol. 38, pp. 1329–1340

[DiBaKaZi92b] DIABY, M., BAHL, H.C., KARWAN, M.H., ZIONTS, S., (1992), Capacitated Lot–Sizing and Scheduling by Lagrangean Relaxation, European Journal of Operational Research, Vol. 59, pp. 444–458

[DiEsWoZh93] DILLENBERGER, C., ESCUDERO, L.F., WOLLSENSAK, A., ZHANG, W., (1993), On Solving a Large–Scale Resource Allocation Problem in Production Planning, in: Fandel, G., Gulledge, Th., Jones, A., Operations Research in Production Planning and Control, Berlin, Springer, pp. 105–119

[DiRa89] Dilts, D.M., Ramsing, K.D., (1989), Joint Lot Sizing and Scheduling of Multiple Items with Sequence Dependent Setup Costs, Decision Sciences, Vol. 20, pp. 120–133

[Din64] Dinkelbach, W., (1964), Zum Problem der Produktionsplanung in Ein– und Mehrproduktunternehmen, Würzburg, Physica, 2nd edition

[Din82] Dinkelbach, W., (1982), Entscheidungsmodelle, Berlin, de Gruyter

[DiSi81] Dixon, P.S., Silver, E.A., (1981), A Heuristic Solution Procedure for the Multi–Item, Single–Level, Limited Capacity Lot–Sizing Problem, Journal of Operations Management, Vol. 2, pp. 23–39

[Dob92] Dobson, G., (1992), The Cyclic Lot Scheduling Problem with Sequence–Dependent Setups, Operations Research, Vol. 40, pp. 736–749

[DoPa77] Doersch, R.H., Patterson, J.H., (1977), Scheduling a Project to Maximize its Present Value: A Zero–One Programming Approach, Management Science, Vol. 23, pp. 882–889

[DoDr91] Domschke, W., Drexl, A., (1991), Kapazitätsplanung in Netzwerken — Ein Überblick über neuere Modelle und Verfahren, OR Spektrum, Vol. 13, pp. 63–76

[DoDr95] Domschke, W., Drexl, A., (1995), Einführung in Operations Research, Berlin, Springer, 3rd edition

[DoScVo93] Domschke, W., Scholl, A., Voss, S., (1993), Produktionsplanung – Ablauforganisatorische Aspekte, Berlin, Springer

[DoPe95] Dorndorf, U., Pesch, E., (1995), Evolution Based Learning in a Job Shop Scheduling Environment, Computers & Operations Research, Vol. 22, pp. 25–40

[Dre91] Drexl, A., (1991), Scheduling of Project Networks by Job Assignment, Management Science, Vol. 37, pp. 1590–1602

[DrFlGüStTe94] Drexl, A., Fleischmann, B., Günther, H.O., Stadtler, H., Tempelmeier, H., (1994), Konzeptionelle Grundlagen kapazitätsorientierter PPS–Systeme, Zeitschrift für betriebswirtschaftliche Forschung, Vol. 46, pp. 1022–1045

[DrGr93] Drexl, A., Grünewald, J., (1993), Nonpreemptive Muli–Mode Resource–Constrained Project Scheduling, IIE Transactions, Vol. 25, No. 5, pp. 74–81

[DrHa95] DREXL, A., HAASE, K., (1995), Proportional Lotsizing and Scheduling, International Journal of Production Economics, Vol. 40, pp. 73–87

[DrHa96] DREXL, A., HAASE, K., (1996), Sequential–Analysis Based Randomized–Regret–Methods for Lot–Sizing and Scheduling, Journal of the Operational Research Society, Vol. 47, pp. 251–265

[DrHaKi95] DREXL, A., HAASE, K., KIMMS, A., (1995), Losgrößen– und Ablaufplanung in PPS–Systemen auf der Basis randomisierter Opportunitätskosten, Zeitschrift für Betriebswirtschaft, Vol. 65, pp. 267–285

[DrJuSa93] DREXL, A., JURETZKA, J., SALEWSKI, F., (1993), Academic Course Scheduling under Workload and Changeover Constraints, Working Paper, No. 337, University of Kiel

[DrGuPa95] DREZNER, Z., GURNANI, H., PASTERNACK, B.A., (1995), An EOQ Model with Substitutions between Products, Journal of the Operational Research Society, Vol. 46, pp. 887–891

[Dyc90] DYCKHOFF, H., (1990), A Typology of Cutting and Packing Problems, European Journal of Operational Research, Vol. 44, pp. 145–159

[DzGo65] DZIELINSKI, B.P., GOMORY, R.E., (1965), Optimal Programming of Lot Sizes, Inventory and Labour Allocation, Management Science, Vol. 11, pp. 874–890

[Eif75] VON EIFF, W., (1975), Fallstudien aus der Unternehmenspraxis III — Produktionsplanung, Hamburg, Deutsche Unilever GmbH

[EiWe93] EISENFÜHR, F., WEBER, M., (1993), Rationales Entscheiden, Berlin, Springer

[Elm63] ELMAGHRABY, S.E., (1963), A Note on the 'Explosion' and 'Netting' Problems in the Planning of Materials Requirements, Operations Research, Vol. 11, pp. 530–535

[Elm78] ELMAGHRABY, S.E., (1978), The Economic Lot Scheduling Problem (ELSP): Review and Extensions, Management Science, Vol. 24, pp. 587–598

[ElGi64] ELMAGHRABY, S.E., GINSBERG, A.S., (1964), A Dynamic Model for Optimal Loading of Linear Multi–Operation Shops, Management Technology, Vol. 4, pp. 47–58

[ENa89] EL–NAJDAWI, M.K., (1989), Common Cycle Scheduling Approach to Lot–Size Scheduling for Multi–Stage, Multi–Product Production Processes, Ph.D. dissertation, University of Pennsylvania

[ENa92] EL–NAJDAWI, M.K., (1992), A Compact Heuristic for Lot–Size Scheduling in Multi–Stage, Multi–Product Production Processes, International Journal of Production Economics, Vol. 27, pp. 29–41

[ENa94] EL–NAJDAWI, M.K., (1994), A Job–Splitting Heuristic for Lot–Size Scheduling in Multi–Stage, Multi–Product Production Processes, European Journal of Operational Research, Vol. 75, pp. 365–377

[ENaKl93] EL–NAJDAWI, M.K., KLEINDORFER, P.R., (1993), Common Cycle Lot–Size Scheduling for Multi–Product, Multi–Stage Production, Management Science, Vol. 39, pp. 872–885

[EpGoPa69] EPPEN, G.D., GOULD, F.J., PASHIGIAN, B.P., (1969), Extensions of Planning Horizon Theorems in the Dynamic Lot Size Model, Management Science, Vol. 15, pp. 268–277

[EpMa87] EPPEN, G.D., MARTIN, R.K., (1987), Solving Multi–Item Capacitated Lot–Sizing Problems Using Variable Redefinition, Operations Research, Vol. 35, pp. 832–848

[Ere88] ERENGUC, S.S., (1988), Multiproduct Dynamic Lot–Sizing Model with Coordinated Replenishments, Naval Research Logistics, Vol. 35, pp. 1–22

[FaKe92] FAIGLE, U., KERN, W., (1992), Some Convergence Results for Probabilistic Tabu Search, ORSA Journal on Computing, Vol. 4, pp. 32–37

[FeTz91] FEDERGRUEN, A., TZUR, M., (1991), A Simple Forward Algorithm to Solve General Dynamic Lot Sizing Models with n Periods in $O(n \log n)$ or $O(n)$ Time, Management Science, Vol. 37, pp. 909–925

[FeTz93] FEDERGRUEN, A., TZUR, M., (1993), The Dynamic Lot–Sizing Model with Backlogging: A Simple $O(n \log n)$ Algorithm and Minimal Forecast Horizon Procedure, Naval Research Logistics, Vol. 40, pp. 459–478

[FeTz94a] FEDERGRUEN, A., TZUR, M., (1994), Minimal Forecast Horizons and a New Planning Procedure for the General Dynamic Lot Sizing Model: Nervousness Revisited, Operations Research, Vol. 42, pp. 456–468

[FeTz94b] FEDERGRUEN, A., TZUR, M., (1994), The Joint Replenishment Problem with Time–Varying Costs and Demands: Efficient, Asymptotic and ϵ–Optimal Solutions, Operations Research, Vol. 42, pp. 1067–1086

[FeTz95] FEDERGRUEN, A., TZUR, M., (1995), Fast Solution and Detection of Minimal Forecast Horizons in Dynamic Programs with a Single Indicator of the Future: Applications to Dynamic Lot–Sizing Models, Management Science, Vol. 41, pp. 874–893

[FeZh92] FEDERGRUEN, A., ZHENG, Y.S., (1992), The Joint Replenishment Problem with General Joint Cost Structures, Operations Research, Vol. 40, pp. 384–403

[FeReSm94] FEO, T.A., RESENDE, M.G.C., SMITH, S.H., (1994), A Greedy Randomized Adaptive Search Procedure for Maximum Independent Set, Operations Research, Vol. 42, pp. 860–878

[Fis81] FISHER, M.L., (1981), The Lagrangian Relaxation Method for Solving Integer Programming Problems, Management Science, Vol. 27, pp. 1–18

[Fis85] FISHER, M.L., (1985), An Applications Oriented Guide to Lagrangian Relaxation, Interfaces, Vol. 15, No. 2, pp. 10–21

[Fle88] FLEISCHMANN, B., (1988), Operations–Research–Modelle und –Verfahren in der Produktionsplanung, Zeitschrift für betriebswirtschaftliche Forschung, Vol. 58, pp. 347–372

[Fle90] FLEISCHMANN, B., (1990), The Discrete Lot–Sizing and Scheduling Problem, European Journal of Operational Research, Vol. 44, pp. 337–348

[Fle94] FLEISCHMANN, B., (1994), The Discrete Lot–Sizing and Scheduling Problem with Sequence–Dependent Setup Costs, European Journal of Operational Research, Vol. 75, pp. 395–404

[FlMe96] FLEISCHMANN, B., MEYR, H., (1996), The General Lot–Sizing and Sequencing Problem, Working Paper, University of Augsburg

[Fre88] FRENCH, S., (1988), Decision Theory: An Introduction to the Mathematics of Rationality, Chichester, Ellis Horwood/John Wiley

[Fuk88] FUKUSHIMA, K., (1988), A Neural Network for Visual Pattern Recognition, IEEE Computer, Vol. 3, pp. 65–75

[Gab79] GABBAY, H., (1979), Multi–Stage Production Planning, Management Science, Vol. 25, pp. 1138–1149

[GaJo94] GALLEGO, G., JONEJA, D., (1994), Economic Lot Scheduling Problem with Raw Material Considerations, Operations Research, Vol. 42, pp. 92–101

[Gar70] GARDNER, M., (1970), The Fantastic Combinations of Conway's New Solitaire Game "Life", Scientific American, Vol. 223, pp. 120–123

[GaJo79] GAREY, M.R., JOHNSON, D.S., (1979), Computers and Intractability: A Guide to the Theory of NP–completeness, San Francisco, Freeman

[Geo74] GEOFFRION, A.M., (1974), Lagrangian Relaxation and its Uses in Integer Programming, Mathematical Programming Study, Vol. 2, pp. 82–114

[Gil88] GILLESSEN, E., (1988), Integrierte Produktionsplanung: Lagerhaltung und Fremdbezug als Bestandteil eines ganzheitlichen Planungskonzeptes, Berlin, Springer

[Glo89] GLOVER, F., (1989), Tabu Search — Part I, ORSA Journal on Computing, Vol. 1, pp. 190–206

[Glo90a] GLOVER, F., (1990), Tabu Search — Part II, ORSA Journal on Computing, Vol. 2, pp. 4–32

[Glo90b] GLOVER, F., (1990), Tabu Search: A Tutorial, Interfaces, Vol. 20, No. 4, pp. 74–94

[Glo94] GLOVER, F., (1994), Tabu Search Fundamentals and Uses, Working Paper, University of Boulder

[Gol89] GOLDBERG, D.E., (1989), Genetic Algorithms in Search, Optimization, and Machine Learning, Reading, Addison–Wesley

[GoLeGaXi94] GONÇALVES, J.F., LEACHMAN, R.C., GASCON, A., XIONG, Z.K., (1994), Heuristic Scheduling Policy for Multi–Item, Multi–Machine Production Systems with Time–Varying, Stochastic Demands, Management Science, Vol. 40, pp. 1455–1468

[Gor70] GORENSTEIN, S., (1970), Planning Tire Production, Management Science, Vol. 17, pp. B72–B82

[Goy76] GOYAL, S.K., (1976), Note on "Manufacturing Cycle Time Determination for a Multi–Stage Economic Production Quantity Model", Management Science, Vol. 23, pp. 332–333

[GoGu90] GOYAL, S.K., GUNASEKARAN, A., (1990), Multi–Stage Production–Inventory Systems, European Journal of Operational Research, Vol. 46, pp. 1–20

[GrLaLeRKa79] GRAHAM, R.L., LAWLER, E.L., LENSTRA, J.K., RINNOOY KAN, A.H.G., (1979), Optimization and Approximation in Deterministic Sequencing and Scheduling: A Survey, Annals of Discrete Mathematics, Vol. 5, pp. 287–326

[Gra81] GRAVES, S.C., (1981), Multi–Stage Lot–Sizing: An Iterative Procedure, in: Schwarz, L.B., (ed.) Multi–Level Production/Inventory Control Systems: Theory and Practice, Studies in the Management Science, Vol. 16, Amsterdam, North–Holland, pp. 95–109

[GrMeDaQu86] GRAVES, S.C., MEAL, H.C., DASU, S.D., QUI, Y., (1986), Two–Stage Production Planning in a Dynamic Environment, in: Axsäter, S., Schneeweiss, C., Silver, E., (eds.), Multi–Stage Production Planning and Inventory Control, Berlin, Springer, pp. 9–43

[GrSc77] GRAVES, S.C., SCHWARZ, L.B., (1977), Single Cycle Continuous Review Policies for Arborescent Production/Inventory Systems, Management Science, Vol. 23, pp. 529–540

[GrSc78] GRAVES, S.C., SCHWARZ, L.B., (1978), On Stationarity and Optimality in Arborescent Production/Inventory Systems, Management Science, Vol. 24, pp. 1768–1769

[Gün86] GÜNTHER, H.O., (1986), The Design of a Hierarchical Model for Production Planning and Scheduling, in: Axsäter, S., Schneeweiß, C., Silver, E., (eds.), Multi–Stage Production Planning and Inventory Control, Berlin, Springer, pp. 227–260

[Gün87] GÜNTHER, H.O., (1987), Planning Lot Sizes and Capacity Requirements in a Single–Stage Production System, European Journal of Operational Research, Vol. 31, pp. 223–231

[GuBr92] GUPTA, S.M., BRENNAN, L., (1992), Lot Sizing and Backordering in Multi–Level Product Structures, Production and Inventory Management Journal, Vol. 33, No. 1, pp. 27–35

[GuKe90] GUPTA, Y.P., KEUNG, Y.K., (1990), A Review of Multi–Stage Lot–Sizing Models, International Journal of Production Management, Vol. 10, pp. 57–73

[GuKeGu92] GUPTA, Y.P., KEUNG, Y.K., GUPTA, M.C., (1992), Comparative Analysis of Lot-Sizing Models for Multi-Stage Systems: A Simulation Study, International Journal of Production Research, Vol. 30, pp. 695–716

[Haa93] HAASE, K., (1993), Capacitated Lot-Sizing with Linked Production Quantities of Adjacent Periods, Working Paper No. 334, University of Kiel

[Haa94] HAASE, K., (1994), Lotsizing and Scheduling for Production Planning, Lecture Notes in Economics and Mathematical Systems, Vol. 408, Berlin, Springer

[Haa96] HAASE, K., (1996), Capacitated Lot-Sizing with Sequence Dependent Setup Costs, OR Spektrum, Vol. 18, pp. 51–59

[HaKi96] HAASE, K., KIMMS, A., (1996), Lot Sizing and Scheduling with Sequence Dependent Setup Costs and Times and Efficient Rescheduling Opportunities, Working Paper No. 393, University of Kiel

[HvL67] HAEHLING VON LANZENAUER, C., (1967), Optimale Lagerhaltung bei mehrstufigen Produktionsprozessen, Unternehmensforschung, Vol. 11, pp. 33–48

[HvL70a] HAEHLING VON LANZENAUER, C., (1970), A Production Scheduling Model by Bivalent Linear Programming, Management Science, Vol. 17, pp. 105–111

[HvL70b] HAEHLING VON LANZENAUER, C., (1970), Production and Employment Scheduling in Multistage Production Systems, Naval Research Logistics, Vol. 17, pp. 193–198

[HaYa95a] HAHM, J., YANO, C.A., (1995), The Economic Lot and Delivery Scheduling Problem: The Common Cycle Case, IIE Transactions, Vol. 27, pp. 113–125

[HaYa95b] HAHM, J., YANO, C.A., (1995), The Economic Lot and Delivery Scheduling Problem: Models for Nested Schedules, IIE Transactions, Vol. 27, pp. 126–139

[Hal88] HALL, N.G., (1988), A Multi-Item EOQ with Inventory Cycle Balancing, Naval Research Logistics, Vol. 35, pp. 319–325

[HaJa90] HANSEN, P., JAUMARD, B., (1990), Algorithms for the Maximum Satisfiability Problem, Computing, Vol. 44, pp. 279–303

[HaJa95] HARIGA, M.A., JACKSON, P.L., (1995), Time-Variant Lot Sizing Models for the Warehouse Scheduling Problem, IIE Transactions, Vol. 27, pp. 162–170

[HaRi85] HARL, J.E., RITZMAN, L.P., (1985), An Algorithm to Smooth Near–Term Capacity Requirements Generated by MRP Systems, Journal of Operations Management, Vol. 5, pp. 309–326

[Har13] HARRIS, F.W., (1913), How Many Parts to Make at Once, Factory, The Magazine of Management, Vol. 10, pp. 135–136, 152, reprinted (1990), Operations Research, Vol. 38, pp. 947–950

[HaCl85] HAYES, R., CLARK, K., (1985), Explaining Observed Productivity Differentials between Plants: Implications for Operations Research, Decision Sciences, Vol. 16, pp. 3–14

[Hec91] HECHTFISCHER, R., (1991), Kapazitätsorientierte Verfahren der Losgrößenplanung, Wiesbaden, Deutscher Universitäts–Verlag

[Hei87] HEINRICH, C.E., (1987), Mehrstufige Losgrößenplanung in hierarchisch strukturierten Produktionsplanungssystemen, Berlin, Springer

[HeSc86] HEINRICH, C.E., SCHNEEWEISS, C., (1986), Multi–Stage Lot–Sizing for General Production Systems, in: Axsäter, S., Schneeweiß, C., Silver, E., (eds.), Multi–Stage Production Planning and Inventory Control, Berlin, Springer, pp. 150–181

[Hel94] HELBER, S., (1994), Kapazitätsorientierte Losgrößenplanung in PPS–Systemen, Stuttgart, M & P

[Hel95] HELBER, S., (1995), Lot Sizing in Capacitated Production Planning and Control Systems, OR Spektrum, Vol. 17, pp. 5–18

[HeWoCr74] HELD, M., WOLFE, P., CROWDER, H.P., (1974), Validation of Subgradient Optimization, Mathematical Programming, Vol. 6, pp. 62–88

[Her72] HERROELEN, W.S., (1972), Resource–Constrained Project Scheduling — The State of the Art, Operational Research Quarterly, Vol. 23, pp. 261–275

[HeWe91] HERTZ, A., DE WERRA, D., (1991), The Tabu Search Metaheuristic: How We Used It, Annals of Mathematics and Artificial Intelligence, Vol. 1, pp. 111–121

[HiLi88] HILLIER, F.S., LIEBERMANN, G.J., (1988), Operations Research — Einführung, München, Oldenbourg, 4th edition

[Hin95] HINDI, K.S., (1995), Algorithms for Capacitated, Multi–Item Lot–Sizing without Set–ups, Journal of the Operational Research Society, Vol. 46, pp. 465–472

[Hin96] HINDI, K.S., (1996), Solving the CLSP by a Tabu Search Heuristic, Journal of the Operational Research Society, Vol. 47, pp. 151–161

[Ho93] HO, C., (1993), Evaluating Lot–Sizing Performance in Multi–Level MRP–Systems: A Comparative Analysis of Multiple Performance Measures, International Journal of Operations and Production Management, Vol. 13, pp. 52–79

[HoKo94] VAN HOESEL, S., KOLEN, A., (1994), A Linear Description of the Discrete Lot–Sizing and Scheduling Problem, European Journal of Operational Research, Vol. 75, pp. 342–353

[Hof94] HOFMANN, C., (1994), Abstimmung von Produktions– und Transportlosgrößen zwischen Zulieferer und Produzent — Eine Analyse auf der Grundlage stationärer Losgrößenmodelle, OR Spektrum, Vol. 16, pp. 9–20

[Hol75] HOLLAND, J.H., (1975), Adaptation in Natural and Artificial Systems, Ann Arbor, The University of Michigan Press

[Hol79] HOLLOWAY, C.A., (1979), Decision Making under Uncertainty: Models and Choices, Englewood Cliffs, Prentice Hall

[Hom95] HOMBURG, C., (1995), Single Sourcing, Double Sourcing, Multiple Sourcing,...?, Zeitschrift für Betriebswirtschaft, Vol. 65, pp. 813–834

[Hoo95] HOOKER, J.N., (1995), Testing Heuristics: We Have It All Wrong, Journal of Heuristics, Vol. 1, pp. 33–42

[HoSa76] HOROWITZ, E. SAHNI, S., (1976), Exact and Approximate Algorithms for Scheduling Nonidentical Processors, Journal of the ACM, Vol. 23, pp. 317–327

[HsLaCh92] HSIEH, H., LAM, K.F., CHOO, E.U., (1992), Comparative Study of Dynamic Lot Sizing Heuristics with Backlogging, Computers & Operations Research, Vol. 19, pp. 393–407

[Hsu84] HSU, J.I.S., (1984), Economic Production Quantity Determination in the Multistage Process, Production and Inventory Management Journal, Vol. 25, No. 4, pp. 61–70

[HsENa90] HSU, J.I.S., EL–NAJDAWI, M.K., (1990), Common Cycle Scheduling in a Multistage Production Process, Engineering Costs and Production Economics, Vol. 20, pp. 73–80

[HuSa91] HUM, S.H., SARIN, R.K., (1991), Simultaneous Product–Mix Planning, Lot Sizing and Scheduling at Bottleneck Facilities, Operations Research, Vol. 39, pp. 296–307

[HuRoWa94] HUNG, M.S., ROM, W.O., WAREN, A.D., (1994), Optimization with IBM's OSL, Danvers, boyd & fraser

[IcErZa93] ICMELI, O., ERENGUC, S.S., ZAPPE, J.C., (1993), Project Scheduling Problems: A Survey, International Journal of Operations and Production Management, Vol. 13, pp. 80–91

[Ind95] INDERFURTH, K., (1995), Multistage Safety Stock Planning with Item Demands Correlated Across Products and Through Time, Production and Operations Management, Vol. 4, pp. 127–144

[IyAt93] IYOGUN, P., ATKINS, D., (1993), A Lower Bound and an Efficient Heuristic for Multistage Multiproduct Distribution Systems, Management Science, Vol. 39, pp. 204–217

[JaMaMu85] JACKSON, P.L., MAXWELL, W.L., MUCKSTADT, J.A., (1985), The Joint Replenishment Problem with a Powers of Two Restriction, IIE Transactions, Vol. 17, pp. 25–32

[JaMaMu88] JACKSON, P.L., MAXWELL, W.L., MUCKSTADT, J.A., (1988), Determining Optimal Reorder Intervals in Capacitated Production–Distribution Systems, Management Science, Vol. 34, pp. 938–958

[JaKh80] JACOBS, F.R., KHUMAWALA, B.M., (1980), Multi–Level Lot Sizing: An Experimental Analysis, Proceedings of the American Institute of Decision Sciences, Atlanta, GA, p. 288

[JaKh82] JACOBS, F.R., KHUMAWALA, B.M., (1982), Multi–Level Lot Sizing in Material Requirements Planning: An Empirical Investigation, Computers & Operations Research, Vol. 9, pp. 139–144

[Jah95] JAHNKE, H., (1995), Produktion bei Unsicherheit — Elemente einer betriebswirtschaftlichen Produktionslehre bei Unsicherheit, Physica–Schriften zur Betriebswirtschaft, Vol. 50, Heidelberg, Physica

[JeKa72] JENSEN, P., KAHN, H.A., (1972), Scheduling in a Multi-Stage Production System with Set-Up and Inventory Costs, AIIE Transactions, Vol. 4, pp. 126–133

[Jen93] JENSEN, T., (1993), Nervousness and Reorder Policies in Rolling Horizon Environments, in: Fandel, G., Gulledge, Th., Jones, A., Operations Research in Production Planning and Control, Berlin, Springer, pp. 428–443

[Jon90] JONEJA, D., (1990), The Joint Replenishment Problem: New Heuristics and Worst Case Performance Bounds, Operations Research, Vol. 38, pp. 711–723

[Jon91] JONEJA, D., (1991), Multi-Echelon Assembly Systems with Non-Stationary Demands: Heuristics and Worst Case Performance Bounds, Operations Research, Vol. 39, pp. 512–518

[Jor95] JORDAN, C., (1995), Batching and Scheduling — Models and Methods for Several Problem Classes, Ph.D. dissertation, University of Kiel

[Kal72] KALYMON, B.A., (1972), A Decomposition Algorithm for Arborescence Inventory Systems, Operations Research, Vol. 20, pp. 860–874

[Kao79] KAO, E.P.C., (1979), A Multiproduct Dynamic Lot-Size Model with Individual and Joint Setup Costs, Operations Research, Vol. 27, pp. 279–289

[KaKeKe87] KARMARKAR, U.S., KEKRE, S., KEKRE, S., (1987), The Deterministic Lotsizing Problem with Startup and Reservation Costs, Operations Research, Vol. 35, pp. 389–398

[KaKeKeFr85] KARMARKAR, U.S., KEKRE, S., KEKRE, S., FREEMAN, S., (1985), Lot-Sizing and Lead-Time Performance in a Manufacturing Cell, Interfaces, Vol. 15, No. 2, pp. 1–9

[KaSc85] KARMARKAR, U.S., SCHRAGE, L., (1985), The Deterministic Dynamic Product Cycling Problem, Operations Research, Vol. 33, pp. 326–345

[KaMaMo88] KARWAN, K.R., MAZZOLA, J.B., MOREY, R.C., (1988), Production Lot Sizing under Setup and Worker Learning, Naval Research Logistics, Vol. 35, pp. 159–175

[Kim93] KIMMS, A., (1993), A Cellular Automaton Based Heuristic for Multi-Level Lot Sizing and Scheduling, Working Paper No. 331, University of Kiel

[Kim94a] KIMMS, A., (1994), Competitive Methods for Multi-Level Lot Sizing and Scheduling: Tabu Search and Randomized Regrets, Working Paper No. 348, University of Kiel, International Journal of Production Research, to appear

[Kim94b] KIMMS, A., (1994), Optimal Multi-Level Lot Sizing and Scheduling with Dedicated Machines, Working Paper No. 351, University of Kiel

[Kim94c] KIMMS, A., (1994), Demand Shuffle — A Novel Method for Multi-Level Proportional Lot Sizing and Scheduling, Working Paper No. 355, University of Kiel

[Kim95] KIMMS, A., (1995), Zum Problem der taktisch-operativen Entscheidung zwischen Eigen- oder Fremdfertigung, Working Paper No. 374, University of Kiel

[Kim96] KIMMS, A., (1996), Multi-Level, Single-Machine Lot Sizing and Scheduling (with Initial Inventory), European Journal of Operational Research, Vol. 89, pp. 86-99

[Kir95] KIRCA, Ö., (1995), A Primal-Dual Algorithm for the Dynamic Lotsizing Problem with Joint Set-Up Costs, Naval Research Logistics, Vol. 42, pp. 791-806

[KiKö94] KIRCA, Ö., KÖKTEN, M., (1994), A New Heuristic Approach for the Multi-Item Dynamic Lot Sizing Problem, European Journal of Operational Research, Vol. 75, pp. 332-341

[KlKu78] KLEINDORFER, P., KUNREUTHER, H.C., (1978), Stochastic Horizon for the Aggregate Planning Problem, Management Science, Vol. 24, pp. 485-497

[Kol95] KOLISCH, R., (1995), Project Scheduling under Resource Constraints — Efficient Heuristics for Several Problem Classes, Heidelberg, Physica

[KrRiWo80] KRAJEWSKI, L.J., RITZMAN, L.P., WONG, D. S., (1980), The Relationship between Product Structure and Multistage Lot Sizes: A Preliminary Analysis, Proceedings of the American Institute of Decision Sciences, Atlanta, GA, pp. 6-8

[KrCaJu83] KROPP, D.H., CARLSON, R.C., JUCKER, J.V., (1983), Concepts, Theories and Techniques — Heuristic Lot-Sizing Approaches for Dealing with MRP System Nervousness, Decision Sciences, Vol. 14, pp. 156-169

[KuSa90] KUIK, R., SALOMON, M., (1990), Multi–Level Lot–Sizing Problem: Evaluation of a Simulated–Annealing Heuristic, European Journal of Operational Research, Vol. 45, pp. 25–37

[KuSaWaMa93] KUIK, R., SALOMON, M., VAN WASSENHOVE, L.N., MAES, J., (1993), Linear Programming, Simulated Annealing and Tabu Search Heuristics for Lotsizing in Bottleneck Assembly Systems, IIE Transactions, Vol. 25, No. 1, pp. 62–72

[KuSaWa94] KUIK, R., SALOMON, M., VAN WASSENHOVE, L.N., (1994), Batching Decisions: Structure and Models, European Journal of Operational Research, Vol. 75, pp. 243–263

[KuMo73] KUNREUTHER, H.C., MORTON, T.E., (1973), Planning Horizons for Production Smoothing with Deterministic Demands, Management Science, Vol. 20, pp. 110–125

[LFo82] LAFORGE, R.L., (1982), MRP and the Part Period Algorithm, Journal of Purchasing and Materials Management, Vol. 18, pp. 21–26

[LFo85] LAFORGE, R.L., (1985), A Decision Rule for Creating Planned Orders in MRP, Production and Inventory Management Journal, Vol. 26, pp. 115–126

[LaVEe78] LAMBRECHT, M.R., VANDER EECKEN, J., (1978), A Facilities in Series Capacity Constrained Dynamic Lot–Size Model, European Journal of Operational Research, Vol. 2, pp. 42–49

[LaVEeVa81] LAMBRECHT, M.R., VANDER EECKEN, J., VANDERVEKEN, H., (1981), Review of Optimal and Heuristic Methods for a Class of Facilities in Series Dynamic Lot–Size Problems, in: Schwarz, L.B., (ed.), Multi–Level Production / Inventory Control Systems: Theory and Practice, Studies in the Management Science, Vol. 16, Amsterdam, North–Holland, pp. 69–94

[LaVEeVa83] LAMBRECHT, M.R., VANDER EECKEN, J., VANDERVEKEN, H., (1983), A Comparative Study of Lot Sizing Procedures for Multi–Stage Assembly Systems, OR Spektrum, Vol. 5, pp. 33–43

[LaTe71] LASDON, L.S., TERJUNG, R.C., (1971), An Efficient Algorithm for Multi–Item Scheduling, Operations Research, Vol. 19, pp. 946–969

[Las92] LASSERRE, J.B., (1992), An Integrated Model for Job-Shop Planning and Scheduling, Management Science, Vol. 38, pp. 1201–1211

[Lau91] LAUX, H., (1991), Entscheidungstheorie I — Grundlagen, Berlin, Springer, 2nd edition

[Lau93] LAUX, H., (1993), Entscheidungstheorie II — Erweiterung und Vertiefung, Berlin, Springer, 3rd edition

[LeGaXi93] LEACHMAN, R.C., GASCON, A., XIONG, Z.K., (1993), Multi–Item Single–Machine Scheduling with Material Supply Constraints, Journal of the Operational Research Society, Vol. 44, pp. 1145–1154

[LeDe86] LEE, C.Y., DENARDO, E.V., (1986), Rolling Planning Horizons: Error Bounds for the Dynamic Lot Size Model, Mathematics of Operations Research, Vol. 11, pp. 423–432

[LeBi93] LEE, H.L., BILLINGTON, C., (1993), Material Management in Decentralized Supply Chains, Operations Research, Vol. 41, pp. 835–847

[LeBiCa93] LEE, H.L., BILLINGTON, C., CARTER, B., (1993), Hewlett–Packard Gains Control of Inventory and Service through Design for Localization, Interfaces, Vol. 23, No. 4, pp. 1–11

[LeZi89] LEE, S.B., ZIPKIN, P.H., (1989), A Dynamic Lot–Size Model with Make–or–Buy Decisions, Management Science, Vol. 35, pp. 447–458

[LiHi89] LIEPINS, G.E., HILLIARD, M.R., (1989), Genetic Algorithms: Foundations and Applications, Annals of Operations Research, Vol. 21, pp. 31–58

[LINDO93] LINDO SYSTEMS INC., (1993), LINGO — Optimization Modeling Language, Chicago

[LoCh91] LOTFI, V., CHEN, W.H., (1991), An Optimal Algorithm for the Multi–Item Capacitated Production Planning Problem, European Journal of Operational Research, Vol. 52, pp. 179–193

[Lov72] LOVE, S.F., (1972), A Facilities in Series Inventory Model with Nested Schedules, Management Science, Vol. 18, pp. 327–338

[LoLaOn91] LOZANO, S., LARRANETA, J., ONIEVA, L., (1991), Primal–Dual Approach to the Single Level Capacitated Lot–Sizing Problem, European Journal of Operational Research, Vol. 51, pp. 354–366

[LuMo75] LUNDIN, R.A., MORTON, T.E., (1975), Planning Horizons for the Dynamic Lot Size Model: Zabel vs. Protective Procedures and Computational Results, Operations Research, Vol. 33, pp. 711–734

[Män81] MÄNNEL, W., (1981), Die Wahl zwischen Eigenfertigung und Fremdbezug: Theoretische Grundlagen, praktische Fälle, Stuttgart, Poeschel, 2nd edition

[Mae87] MAES, J., (1987), Capacitated Lotsizing Techniques in Manufacturing Resource Planning, Ph.D. dissertation, University of Leuven

[MaMClWa91] MAES, J., MCCLAIN, J.O., VAN WASSENHOVE, L.N., (1991), Multilevel Capacitated Lotsizing Complexity and LP–Based Heuristics, European Journal of Operational Research, Vol. 53, pp. 131–148

[MaWa88] MAES, J, VAN WASSENHOVE, L.N., (1988), Multi–Item Single–Level Capacitated Dynamic Lot–Sizing Heuristics: A General Review, Journal of the Operational Research Society, Vol. 39, pp. 991–1004

[MaWa91] MAES, J., VAN WASSENHOVE, L.N., (1991), Capacitated Dynamic Lot–Sizing Heuristics for Serial Systems, International Journal of Production Research, Vol. 29, pp. 1235–1249

[MaWa93] MALIK, K., WANG, Y., (1993), An Improved Algorithm for the Dynamic Lot–Sizing Problem with Learning Effect in Setups, Naval Research Logistics, Vol. 40, pp. 925–931

[Mar87] MARTIN, R.K., (1987), Generating Alternative Mixed–Integer Programming Models Using Variable Redefinition, Operations Research, Vol. 35, pp. 820–831

[Mat77] MATHER, H., (1977), Reschedule the Reschedules You Just Rescheduled — Way of Life for MRP?, Production and Inventory Management Journal, Vol. 18, pp. 60–79

[Mat91] MATHES, H.D., (1991), Entscheidungshorizonte in der mehrstufigen dynamischen Losgrößenplanung, OR Spektrum, Vol. 13, pp. 77–85

[Mat93] MATHES, H.D., (1993), Some Valid Constraints for the Capacitated Assembly Line Lotsizing Problem, in: Fandel, G., Gulledge, Th., Jones, A., Operations Research in Production Planning and Control, Berlin, Springer, pp. 444–458

[MaGu95] DE MATTA, R., GUIGNARD, M., (1995), The Performance of Rolling Production Schedules in a Process Industry, IIE Transactions, Vol. 27, pp. 564–573

[MaMu85] MAXWELL, W.L., MUCKSTADT, J.A., (1985), Establishing Consistent and Realistic Reorder Intervals in Production–Distribution Systems, Operations Research, Vol. 33, pp. 1316–1341

[MCMMTW82] MCCLAIN, J.O., MAXWELL, W.L., MUCKSTADT, J.A., THOMAS, L.J., WEISS, E.N., (1982), On MRP Lot Sizing, Management Science, Vol. 28, pp. 582–584

[MClTr85] MCCLAIN, J.O., TRIGEIRO, W.W., (1985), Cyclic Assembly Schedules, IIE Transactions, Vol. 17, pp. 346–353

[MKnSaCo91] MCKNEW, M.A., SAYDAM, C., COLEMAN, B. J., (1991), An Efficient Zero–One Formulation of the Multilevel Lot–Sizing Problem, Decision Sciences, Vol. 22, pp. 280–295

[McLa76] MCLAREN, B.J., (1976), A Study of Multiple Level Lot-sizing Procedures for Material Requirements Planning Systems, Ph.D. dissertation, Purdue University

[MiDa86] MINIFIE, R., DAVIS, R., (1986), Survey of MRP Nervousness Issues, Production and Inventory Management Journal, Vol. 27, pp. 111–120

[MoBi77] MOGG, J.M., BIGHAM, P.E., (1977), Scheduling a Multi–Stage Production System with Startup Delays, Omega, Vol. 6, pp. 183–187

[Moi82] MOILY, J.P., (1982), Optimal and Heuristic Lot–Sizing Procedures for Multi–Stage Manufacturing Systems, Ph.D. dissertation, University of Wisconsin

[Moi86] MOILY, J.P., (1986), Optimal and Heuristic Procedures for Component Lot–Splitting in Multi–Stage Manufacturing Systems, Management Science, Vol. 32, pp. 113–125

[MuSi78] MUCKSTADT, J.A., SINGER, H.M., (1978), Comments on "Single Cycle Continuous Review Policies for Arborescent Production/Inventory Systems", Management Science, Vol. 24, pp. 1766–1768

[MüGoKr88] MÜHLENBEIN, H., GORGES–SCHLEUTER, M., KRÄMER, O., (1988), Evolution Algorithms in Combinatorial Optimization, Parallel Computing, Vol. 7, pp. 65–85

[MüMe65] MÜLLER–MERBACH, H., (1965), Optimale Losgrößen bei mehrstufiger Fertigung, Zeitschrift für wirtschaftliche Fertigung, Vol. 60, pp. 113–118

[MuSp83] MUTH, E.J., SPREMANN, K., (1983), Learning Effects in Economic Lot Sizing, Management Science, Vol. 29, pp. 264–269

[NaHaHe95] NAGAR, A., HADDOCK, J., HERAGU, S., (1995), Multiple and Bicriteria Scheduling: A Literature Survey, European Journal of Operational Research, Vol. 81, pp. 88–104

[New74] NEW, C.C., (1974), Lot–Sizing in Multi–Level Requirements Planning Systems, Production and Inventory Management Journal, Vol. 15, No. 4, pp. 57–72

[OlBu85] OLIFF, M.D., BURCH, E.E., (1985), Multiproduct Production Scheduling at Owens–Corning Fiberglas, Interfaces, Vol. 15, No. 5, pp. 25–34

[Ott93] OTT, R.L., (1993), An Introduction to Statistical Methods and Data Analysis, Belmont, Wadsworth, 4th edition

[PaPa76] PAGE, E., PAUL, R.J., (1976), Multi–Product Inventory Situations with One Restriction, Journal of the Operational Research Society, Vol. 27, pp. 815–834

[Pat73] PATTERSON, J.H., (1973), Alternate Methods of Project Scheduling with Limited Resources, Naval Research Logistics, Vol. 20, pp. 767–784

[Pat84] PATTERSON, J.H., (1984), A Comparison of Exact Approaches for Solving the Multiple Constrained Resource, Project Scheduling Problem, Management Science, Vol. 30, pp. 854–867

[Pen85] PENG, K., (1985), Lot–Sizing Heuristics for Multi–Echelon Assembly Systems, Engineering Costs and Production Economics, Vol. 9, pp. 51–57

[PiFr93] PICOT, A., FRANCK, E., (1993), Vertikale Integration, in: Hauschildt, J., Grün, O., (eds.), Ergebnisse empirischer betriebswirtschaftlicher Forschung — Zu einer Realtheorie der Unternehmung, Stuttgart, Schäffer–Poeschel, pp. 179–219

[Pin95] PINEDO, M., (1995), Scheduling Theory, Englewood Cliffs, Prentice Hall

[PoWo91] POCHET, Y., WOLSEY, L.A., (1991), Solving Multi–Item Lot–Sizing Problems Using Strong Cutting Planes, Management Science, Vol. 37, pp. 53–67

[Pop93] POPP, T., (1993), Kapazitätsorientierte dynamische Losgrößen- und Ablaufplanung bei Sortenproduktion, Hamburg, Kovač

[PoWa92] POTTS, C.N., VAN WASSENHOVE, L.N., (1992), Integrating Scheduling with Batching and Lot-Sizing: A Review of Algorithms and Complexity, Journal of the Operational Research Society, Vol. 43, pp. 395–406

[PrCaRa94] PRATSINI, E., CAMM, J.D., RATURI, A.S., (1994), Capacitated Lot Sizing under Setup Learning, European Journal of Operational Research, Vol. 72, pp. 545–557

[Raj92] RAJAGOPALAN, S., (1992), A Note on "An Efficient Zero–One Formulation of the Multilevel Lot–Sizing Problem", Decision Sciences, Vol. 23, pp. 1023–1025

[RaRa83] RAMSEY, T.E., RARDIN, R.R., (1983), Heuristics for Multistage Production Planning Problems, Journal of the Operational Research Society, Vol. 34, pp. 61–70

[Rao81] RAO, V.V., (1981), Optimal Lot Sizing for Acyclic Multi–Stage Production Systems, Ph.D. dissertation, Georgia Institute of Technology

[RaHi88] RATURI, A., S., HILL, A.V., (1988), An Experimental Analysis of Capacity–Sensitive Setup Parameters for MRP Lot Sizing, Decision Sciences, Vol. 19, pp. 782–800

[Ree93] REEVES, C., (1993), Modern Heuristic Techniques for Combinatorial Problems, Oxford, Blackwell

[ReSt82] REHMANI, Q., STEINBERG, E., (1982), Simple Single Pass, Multiple Level Lot Sizing Heuristics, Proceedings of the American Institute of Decision Sciences, Atlanta, GA, pp. 124–126

[RiVö89] RICHTER, K., VÖRÖS, J., (1989), On the Stability Region for Multi–Level Inventory Problems, European Journal of Operational Research, Vol. 41, pp. 169–173

[Ris90] RISTROPH, J.H., (1990), Simulation Ordering Rules for Single Level Material Requirements Planning with a Rolling Horizon and No Forecast Error, Computers & Industrial Engineering, Vol. 19, pp. 155–159

[RoKa91] ROLL, Y., KARNI, R., (1991), Multi–Item, Multi–Level Lot Sizing with an Aggregate Capacity Constraint, European Journal of Operational Research, Vol. 51, pp. 73–87

[Ros86] ROSLING, K., (1986), Optimal Lot–Sizing for Dynamic Assembly Systems, in: Axsäter, S., Schneeweiß, C., Silver, E.A., (eds.), Multi–Stage Production Planning and Inventory Control, Berlin, Springer, pp. 119–131

[Rou86] ROUNDY, R.O., (1986), A 98%–Effective Lot–Sizing Rule for a Multi–Product, Multi–Stage Production/Inventory System, Mathematics of Operations Research, Vol. 11, pp. 699–727

[Rou89] ROUNDY, R.O., (1989), Rounding Off to Powers of Two in Continuous Relaxations of Capacitated Lot Sizing Problems, Management Science, Vol. 35, pp. 1433–1442

[Rou93] ROUNDY, R.O., (1993), Efficient, Effective Lot Sizing for Multistage Production Systems, Operations Research, Vol. 41, pp. 371–385

[Sal88] SALIGER, E., (1988), Betriebswirtschaftliche Entscheidungstheorie, München, Oldenbourg, 2nd edition

[Sal91] SALOMON, M., (1991), Deterministic Lotsizing Models for Production Planning, Lecture Notes in Economics and Mathematical Systems, Vol. 355, Berlin, Springer

[SaKrKuWa91] SALOMON, M., KROON, L.G., KUIK, R., VAN WASSENHOVE, L.N., (1991), Some Extensions of the Discrete Lotsizing and Scheduling Problem, Management Science, Vol. 37, pp. 801–812

[SaKuWa93] SALOMON, M., KUIK, R., VAN WASSENHOVE, L.N., (1993), Statistical Search Methods for Lotsizing Problems, Annals of Operations Research, Vol. 41, pp. 453–468

[SaSoWaDuDa95] SALOMON, M., SOLOMON, M.M., VAN WASSENHOVE, L.N., DUMAS, Y.D., DAUZÈRE–PÉRES, S., (1995), Discrete Lot Sizing and Scheduling with Sequence Dependent Setup Times and Costs, Working Paper, European Journal of Operational Research, to appear

[Sch79] SCHRAGE, L., (1979), A More Portable FORTRAN Random Number Generator, ACM Transactions on Mathematical Software, Vol. 5, pp. 132–138

[Sch68] SCHUSSEL, G., (1968), Job–Shop Lot Release Sizes, Management Science, Vol. 14, pp. B449–B472

[Sch82] SCHRAGE, L., (1982), The Multiproduct Lot Scheduling Problem, in: Dempster, M.A.H., Lenstra, J.K., Rinnooy Kan, A.H.G., (eds.), Deterministic and Stochastic Scheduling, Dordrecht, pp. 233–244

[Sch73] SCHWARZ, L.B., (1973), A Simple Continuous Review Deterministic One–Warehouse N–Retailer Inventory Problem, Management Science, Vol. 19, pp. 555–566

[ScSc75] SCHWARZ, L.B., SCHRAGE, L., (1975), Optimal and System Myopic Policies for Multi–Echelon Production/Inventory Assembly Systems, Management Science, Vol. 21, pp. 1285–1294

[ScSc78] SCHWARZ, L.B., SCHRAGE, L., (1978), On Echelon Holding Costs, Management Science, Vol. 24, pp. 865–866

[SiEr94a] SIMPSON, N.C., ERENGUC, S.S., (1994), Multiple Stage Production Planning Research: History and Opportunities, Working Paper, State University of New York at Buffalo

[SiEr94b] SIMPSON, N.C., ERENGUC, S.S., (1994), Improved Heuristic Methods for Multiple Stage Production Planning, Working Paper, State University of New York at Buffalo

[SmMo85] SMUNT, T.L., MORTON, T.E., (1985), The Effects of Learning on Optimal Lot Sizes: Further Developments on the Single Product Case, IIE Transactions, Vol. 17, pp. 33–37

[SöSc95] SÖHNER, V., SCHNEEWEISS, C., (1995), Hierarchically Integrated Lot Size Optimization, European Journal of Operational Research, Vol. 86, pp. 73–90

[Spr94] SPRECHER, A., (1994), Resource–Constrained Project Scheduling — Exact Methods for the Multi–Mode Case, Lecture Notes in Economics and Mathematical Systems, Vol. 409, Berlin, Springer

[SpKoDr95] SPRECHER, A., KOLISCH, R., DREXL, A., (1995), Semi–Active, Active, and Non–Delay Schedules for the Resource–Constrained Project Scheduling Problem, European Journal of Operational Research, Vol. 80, pp. 94–102

[SrBeUd87] SRIDHARAN, V., BERRY, W.L., UDAYABHANU, V., (1987), Freezing the Master Production Schedule under Rolling Planning Horizons, Management Science, Vol. 33, pp. 1137–1149

[SrBeUd88] SRIDHARAN, V., BERRY, W.L., UDAYABHANU, V., (1988), Measuring Master Production Schedule Stability under Rolling Horizons, Decision Sciences, Vol. 19, pp. 147–166

[Sta88a] STADTLER, H., (1988), Hierarchische Produktionsplanung bei losweiser Fertigung, Heidelberg, Physica

[Sta88] STADTLER, H., (1988), Medium Term Production Planning with Minimum Lotsizes, International Journal of Production Research, Vol. 26, pp. 553–566

[Sta94] STADTLER, H., (1994), Mixed Integer Programming Model Formulations for Dynamic Multi–Item Multi–Level Capacitated Lotsizing, Working Paper, Technical University of Darmstadt

[Sta95] STADTLER, H., (1995), Reformulations of the Shortest Route Model for Dynamic Multi-Item Multi-Level Capacitated Lotsizing, Working Paper, Technical University of Darmstadt

[Ste75] STEELE, D.C., (1975), The Nervous MRP System: How to do Battle, Production and Inventory Management Journal, Vol. 16, pp. 83–89

[StNa80] STEINBERG, E., NAPIER, H.A., (1980), Optimal Multi–Level Lot Sizing for Requirements Planning Systems, Management Science, Vol. 26, pp. 1258–1271

[StNa82] STEINBERG, E., NAPIER, H.A., (1982), On "A Note on MRP Lot Sizing", Management Science, Vol. 28, pp. 585–586

[StWuPa93] STORER, R.H., WU, S.D., PARK, I., (1993), Genetic Algorithms in Problem Space for Sequencing Problems, in: Fandel, G., Gulledge, Th., Jones, A., (eds.), Operations Research in Production Planning and Control, Berlin, Springer, pp. 584–597

[SuHi93] SUM, C.C., HILL, A.V., (1993), A New Framework for Manufacturing Planning and Control Systems, Decision Sciences, Vol. 24, pp. 739–760

[SuPnYa93] SUM, C.C., PNG, D.O.S., YANG, K.K., (1993), Effects of Product Structure Complexity on Multi–Level Lot Sizing, Decision Sciences, Vol. 24, pp. 1135–1156

[Swi89] SWITALSKI, M., (1989), Hierarchische Produktionsplanung — Konzeption und Einsatzbereich, Heidelberg, Physica

[Sze75] SZENDROVITS, A.Z., (1975), Manufacturing Cycle Time Determination for a Multistage Economic Production Quantity Model, Management Science, Vol. 22, pp. 298–308

[Sze76] SZENDROVITS, A.Z., (1976), On the Optimality of Sub–Batch Sizes for a Multi–Stage EPQ Model — A Rejoinder, Management Science, Vol. 23, pp. 334–338

[Sze78] SZENDROVITS, A.Z., (1978), A Comment on "Optimal and System Myopic Policies for Multi–Echelon Production / Inventory Assembly Systems", Management Science, Vol. 24, pp. 863–864

[Sze81] SZENDROVITS, A.Z., (1981), Comments on the Optimality in Optimal and System Myopic Policies for Multi–Echelon Production/Inventory Assembly Systems, Management Science, Vol. 27, pp. 1081–1087

[Sze83] SZENDROVITS, A.Z., (1983), Non–Integer Optimal Lot Size Ratios in Two–Stage Production/Inventory Systems, International Journal of Production Research, Vol. 21, pp. 323–336

[SzDr80] SZENDROVITS, A.Z., DREZNER, Z., (1980), Optimizing Multi–Stage Production with Constant Lot Size and Varying Numbers of Batches, OMEGA The International Journal of Management Science, Vol. 8, pp. 623–629

[TaSk70] TAHA, H.A., SKEITH, R.W., (1970), The Economic Lot Size in Multistage Production Systems, AIIE Transactions, Vol. 2, pp. 157–162

[Tem95] TEMPELMEIER, H., (1995), Material–Logistik — Grundlagen der Bedarfs– und Losgrößenplanung in PPS–Systemen, Berlin, Springer, 3rd edition

[TeDe93] TEMPELMEIER, H., DERSTROFF, M., (1993), Mehrstufige Mehrprodukt–Losgrößenplanung bei beschränkten Ressourcen und genereller Erzeugnisstruktur, OR Spektrum, Vol. 15, pp. 63–73

[TeDe95] TEMPELMEIER, H., DERSTROFF, M., (1995), A Lagrangean–Based Heuristic for Dynamic Multi–Level Multi–Item Constrained Lotsizing with Setup Times, Working Paper (revised version), University of Cologne, Management Science, to appear

[TeHe94] TEMPELMEIER, H., HELBER, S., (1994), A Heuristic for Dynamic Multi–Item Multi–Level Capacitated Lotsizing for General Product Structures, European Journal of Operational Research, Vol. 75, pp. 296–311

[Tho65] THOMPSON, G.L., (1965), On the Parts Requirement Problem, Operations Research, Vol. 13, pp. 453–461

[ToMa88] TOFFOLI, T., MARGOLUS, N., (1988), Cellular Automata Machines, Cambridge, MIT Press

[ToWi92] TOKLU, B., WILSON, J.M., (1992), A Heuristic for Multi-Level Lot-Sizing Problems with a Bottleneck, International Journal of Production Research, Vol. 30, pp. 787–798

[TrThMCl89] TRIGEIRO, W.W., THOMAS, L.J., MCCLAIN, J.O., (1989), Capacitated Lot Sizing with Setup Times, Management Science, Vol. 35, pp. 353–366

[UnKi92] UNAL, A.T., KIRAN, A.S., (1992), Batch Sequencing, IIE Transactions, Vol. 24, No. 4, pp. 73–83

[Vaz58] VAZSONYI, A., (1958), Scientific Programming in Business and Industry, New York, Wiley

[Vei69] VEINOTT, A.F., (1969), Minimum Concave–Cost Solution of Leontief Substitution Models of Multi–Facility Inventory Systems, Operations Research, Vol. 17, pp. 262–291

[VeLFo85] VERAL, E.A., LAFORGE, R.L., (1985), The Performance of a Simple Incremental Lot–Sizing Rule in a Multilevel Inventory Environment, Decision Sciences, Vol. 16, pp. 57–72

[ViMa86] VICKERY, S.K., MARKLAND, R.E., (1986), Multi–Stage Lot–Sizing in a Serial Production System, International Journal of Production Research, Vol. 24, pp. 517–534

[Vil89] VILLA, A., (1989), Decision Architectures for Production Planning in Multi–Stage Multi–Product Manufacturing Systems, Annals of Operations Research, Vol. 17, pp. 51–68

[VöCh92] VÖRÖS, J., CHAND, S., (1992), Improved Lot Sizing Heuristics for Multi–Stage Inventory Models with Backlogging, International Journal of Production Economics, Vol. 28, pp. 283–288

[Vör95] VÖRÖS, J., (1995), Setup Cost Stability Region for the Multi–Level Dynamic Lot Sizing Problem, European Journal of Operational Research, Vol. 87, pp. 132–141

[WaHoKo92] WAGELMANS, A., VAN HOESEL, S., KOLEN, A., (1992), Economic Lot Sizing: An $O(n \log n)$ Algorithm that Runs in Linear Time in the Wagner–Whitin Case, Operations Research, Vol. 40, pp. S145–S156

[WaWh58] WAGNER, H.M., WHITIN, T.M., (1958), Dynamic Version of the Economic Lot Size Model, Management Science, Vol. 5, pp. 89–96

[Web89] WEBSTER, F.M., (1989), A Back-Order Version of the Wagner-Whitin Discrete Demand EOQ Algorithm by Dynamic Programming, Production and Inventory Management Journal, Vol. 30, pp. 1-5

[Wem82] WEMMERLÖV, U., (1982), An Experimental Analysis of the Use of Echelon Holding Costs and Single-Stage Lot-Sizing Procedures in Multi-Stage Production/Inventory Systems, International Journal of Production Management, Vol. 2, pp. 42-54

[WeWh84] WEMMERLÖV, U., WHYBARK, D.C., (1984), Lot-Sizing under Uncertainty in a Rolling Schedule Environment, International Journal of Production Research, Vol. 22, pp. 467-484

[Whi93] WHITLEY, D., (1993), Foundations of Genetic Algorithms 2, Morgan Kaufmann

[Wil81] WILLIAMS J.F., (1981), Heuristic Techniques for Simultaneous Scheduling of Production and Distribution in Multi-Echelon Structures: Theory and Empirical Comparisons, Management Science, Vol. 27, pp. 336-352

[Wil82] WILLIAMS, J.F., (1982), On the Optimality of Integer Lot Size Ratios in Economic Lot Size Determination in Multi-Stage Assembly Systems, Management Science, Vol. 28, pp. 1341-1349

[Win93] WINSTON, W.L., (1993), Operations Research — Applications and Algorithms, Belmont, Duxbury, 3rd edition

[Wol86] WOLFRAM, S., (1986), Theory and Applications of Cellular Automata, World Scientific

[Wol95] WOLSEY, L.A., (1995), Progress with Single-Item Lot-Sizing, European Journal of Operational Research, Vol. 86, pp. 395-401

[WoSp92] WOODRUFF, D.L., SPEARMAN, M.L., (1992), Sequencing and Batching for Two Classes of Jobs with Deadlines and Setup-Times, Production and Operations Management, Vol. 1, pp. 87-102

[WuStCh93] WU, S.D., STORER, R.H., CHANG, P.C., (1993), One-Machine Rescheduling Heuristics with Efficiency and Stability as Criteria, Computers & Operations Research, Vol. 20, pp. 1-14

[YaLe95] YANO, C.A., LEE, H.L., (1995), Lot Sizing with Random Yields: A Review, Operations Research, Vol. 43, pp. 311–333

[Yel76] YELLE, L.E., (1976), Materials Requirements Lot Sizing: A Multi–Level Approach, International Journal of Production Research, Vol. 19, pp. 223–232

[Zab64] ZABEL, E., (1964), Some Generalization of an Inventory Planning Horizon Theorem, Management Science, Vol. 10, pp. 465–471

[ZäAt80] ZÄPFEL, G., ATTMANN, J., (1980), Losgrößenplanung: Lösungsverfahren für den dynamischen Fall bei beschränkten Kapazitäten und mehrstufiger Fertigung, Das Wirtschaftsstudium, Vol. 9, pp. 122–126 (part I) and pp. 174–177 (part II)

[ZäMi93] ZÄPFEL, G., MISSBAUER, H., (1993), New Concepts for Production Planning and Control, European Journal of Operational Research, Vol. 67, pp. 297–320

[ZaThTr84] ZAHORIK, A., THOMAS, L.J., TRIGEIRO, W. W., (1984), Network Programming Models for Production Scheduling in Multi–Stage, Multi–Item Capacitated Systems, Management Science, Vol. 30, pp. 308–325

[ZaEvVa89] ZANAKIS, S.H., EVANS, J.R., VAZACOPOULOS, A.A., (1989), Heuristic Methods and Applications: A Categorized Survey, European Journal of Operational Research, Vol. 43, pp. 88–110

[Zan66] ZANGWILL, W.I., (1966), A Deterministic Multiproduct, Multifacility Production and Inventory Model, Operations Research, Vol. 14, pp. 486–507

[Zan69] ZANGWILL, W.I., (1969), A Backlogging Model and a Multi–Echelon Model of a Dynamic Economic Lot Size Production System — A Network Approach, Management Science, Vol. 15, pp. 506–527

[Zan87] ZANGWILL, W.I., (1987), Eliminating Inventory in a Series Facility Production System, Management Science, Vol. 33, pp. 1150–1164

[Zol77] ZOLLER, K., (1977), Deterministic Multi–Item Inventory Systems with Limited Capacity, Management Science, Vol. 24, pp. 451–455

List of Tables

List of Figures

Index